Handbooks in Radiology

Skeletal Radiology

HANDBOOKS IN RADIOLOGY SERIES

Series Editors

Anne G. Osborn, M.D.
Director of Neuroradiology and Professor of Radiology, University of Utah School of Medicine, Salt Lake City, Utah

David G. Bragg, M.D.
Professor and Chairman, Department of Radiology, University of Utah School of Medicine, Salt Lake City, Utah

---

Nuclear Medicine
Frederick L. Datz, M.D.

Neuroradiology: Skull and Brain
Anne G. Osborn, M.D., H. Ric Harnsberger, M.D., Wendy R. K. Smoker, M.D.

Chest Radiology
Howard Mann, M.D., David G. Bragg, M.D.

Neuroradiology: Head and Neck
H. Ric Harnsberger, M.D., Wendy R. K. Smoker, M.D., Anne G. Osborn, M.D.

Ultrasonography
Donald A. Cubberley, M.D., William J. Zwiebel, M.D.

Angiography
Myron M. Wojtowycz, M.D.

Neuroradiology: Spine and Spinal Cord
Wendy R. K. Smoker, M.D., B. J. Manaster, M.D., Ph.D., Anne G. Osborn, M.D., H. Ric Harnsberger, M.D.

Handbooks in Radiology
Skeletal Radiology

B. J. Manaster, M.D., Ph.D.
*Associate Professor of Radiology and
General Radiology
Director of Musculoskeletal Division
University of Utah School of Medicine
Salt Lake City, Utah*

YEAR BOOK MEDICAL PUBLISHERS, INC.
CHICAGO · LONDON · BOCA RATON

4 5 6 7 8 9 0 PR 93

Library of Congress Cataloging-in-Publication Data

Manaster, B. J.
  Skeletal radiology/B. J. Manaster.
    p. cm. — (Handbooks in radiology)
  Includes bibliographies and index.
  ISBN 0-8151-5754-1
  1. Skeleton—Diseases—Diagnosis—Handbooks,
  manuals, etc.   2. Skeleton—Radiography—
  Handbooks, manuals, etc. I. Title.
  II. Series.
  [DNLM: 1. Bone and Bones—radiography—
  handbooks.   WE 39 M267s]
  RC930.5.M36   1989
  616.7'10757—dc19
  DNLM/DLC
  for Library of Congress

Sponsoring Editor: James D. Ryan
Associate Managing Editor, Manuscript Services:
  Deborah Thorp
Production Project Coordinator: Carol Reynolds
Proofroom Supervisor: Barbara M. Kelly

To Steve, Tracy Joy, and Katy Rose

Preface

This book is designed primarily for use by diagnostic radiology residents, who will find it a handy source in day-to-day film interpretation, as well as a complete source for board review. Orthopedic residents and clinicians in rheumatology and rehabilitative medicine will find pertinent information relating to their interests.

The book is divided into major sections covering tumor, arthritis, trauma, metabolic bone disease, congenital abnormalities, and a few miscellaneous items. Outline form has been used to allow a quick review. The tumor and arthritis sections begin with an introductory chapter outlining the author's approach to the work-up and diagnosis of these disease processes. The introduction is followed by discussions of the individual diseases, each of which is outlined using the same format; with repeated use, the reader will find that this organization allows easy reference.

A "key concepts" box is found at the beginning of each section. This, by its nature, cannot be all-inclusive, but does give quick reference to the most common features of each disease process.

B. J. Manaster, M.D., Ph.D.

Editors' Introduction

The third text in the *Handbooks in Radiology* series, *Skeletal Radiology*, was written by B.J. Manaster, Associate Professor of Radiology and Director of the Musculoskeletal Section in Radiology at the University of Utah School of Medicine. We believe this is the most complete, functional, and best-organized review of musculoskeletal radiology available, and certainly the most "readable" at the best price. It is intended as an introduction for the medical student and beginning radiologist, yet should find equal application for the practicing radiologist as a review and reference for troubling musculoskeletal cases. It is also a superb review for the resident preparing for National Board Exams.

Skeletal Radiology is subdivided into six major chapters—tumors, arthritis, trauma, metabolic diseases, congenital anomalies, and a section on miscellaneous disorders that covers those entities not easily catalogued by the preceding five broad categories. Illustrations and references are carefully focused to amplify the written message. Each section is complemented by the selected references that include both articles and the best texts relating to each subject in question. The applications and limitations of imaging as applied to musculoskeletal abnormalities are interwoven with radiologic descriptions; an outline format is used throughout to concentrate the material and limit the length of the book. The addition of "key concepts" at the beginning of the description of each disorder provides an "at-a-glance" review of the most pertinent information.

We hope that you will find *Skeletal Radiology* useful, both in its role as an introductory text as well as a review or reference source for the practicing clinician interested in musculoskeletal diseases.

David G. Bragg, M.D.
Anne G. Osborn, M.D.

Contents

Preface **vii**

Editors' Introduction **ix**

CHAPTER 1 TUMOR 1

Generalizations 1
Bone-Forming Tumors 17
Cartilage-Forming Tumors 40
Giant Cell Tumor (GCT) 58
Marrow Tumors 62
Vascular Tumors 71
Other Connective Tissue Tumors 78
Other Tumors 86
Tumorlike Lesions 90
Metastatic Disease of Bone 102

CHAPTER 2 ARTHRITIS 107

Generalizations 107
Polyarthritis of Unknown Etiology 110
Connective Tissue Disorders 139
Rheumatic Fever 147

Osteoarthritis (OA) 149
Neuropathic (Charcot) Joints 158
Biochemical Abnormalities 161
Miscellaneous Disorders 175

CHAPTER 3 TRAUMA 190

Generalizations 190
Fracture Terminology 191
Hand Trauma 196
Wrist Trauma 203
Elbow Trauma 218
Shoulder Trauma 224
Pelvic Trauma 233
Hip Trauma 239
Knee Trauma 247
Ankle Trauma 254
Foot Trauma 260
Spine Trauma 264
Nonaccidental Trauma (Battered Child Syndrome) 279
Myositis Ossificans 281
Orthopedic Hardware 285

CHAPTER 4 METABOLIC BONE DISEASE 293

Osteoporosis 293
Rickets/Osteomalacia 298
Hyperparathyroidism (HPTH) 302
Renal Osteodystrophy 306

Hypoparathyroidism   309

Thyroid Disease   310

Acromegaly   311

Scurvy (Hypovitaminosis C)   312

Mastocytosis   312

Gaucher's Disease   313

Myelofibrosis   314

Paget's Disease of Bone   315

Drug- and Environmentally Induced
   Abnormalities of Bone   317

Serum Laboratory Values in Metabolic Bone
   Disease   319

CHAPTER 5   CONGENTIAL
ANOMALIES   320

Scoliosis   320

Arthrogryposis Multiplex Congenita   324

Neurofibromatosis   325

Marfan's Syndrome   326

Homocystinuria   328

Ehlers-Danlos Syndrome   328

Osteogenesis Imperfecta   329

Sclerosing Dysplasias   331

Congenital Dislocated Hip (CDH)   332

Proximal Femoral Focal Deficiency   336

Infantile Coxa Vara   337

Primary Protrusio of the Acetabulum   337

Pseudarthrosis   338

Congenital Foot Anomalies   338

Tarsal Coalition (Peroneal Spastic
  Flatfoot) 349

Cleidocranial Dysostosis 354

Osteo-Onychodysostosis (Nail-Patella
  Syndrome, or Fong's Disease) 354

Caudal Regression Syndrome (Sacral
  Agenesis) 354

Madelung's Deformity 355

Chromosome Disorders 355

Dwarfism 356
  by Paula Rand

Mucopolysaccharidoses 373

CHAPTER 6 MISCELLANEOUS,
INCLUDING HEMATOLOGIC DISORDERS
AND INFECTION 378

Hematologic Disorders 378

Infection 382

Sarcoidosis 393

Radiation-Induced Abnormalities 394

Acro-osteolysis 395

Periosteal Reaction in Infants 396

Localized Giantism 398

Soft Tissue Calcification Gamut 399

INDEX 402

1

Tumors

GENERALIZATIONS

The individual tumors are discussed according to a slightly modified World Health Organization (WHO) classification in this chapter. The reader is strongly urged to read this introductory chapter first.

A. Purpose of learning the characteristics of bone tumors is to be an effective consultant to the clinician in terms of:
  1. Work-up of a new lesion: Identify a lesion as belonging to one of the following categories:
     a. A benign, "leave me alone" lesion that can best be totally ignored (most notable example, fibrous cortical defect).
     b. A lesion that is almost certainly benign and can safely be watched for confirmation of the diagnosis (examples might include nonossifying fibroma, fibrous dysplasia, and myositis ossificans).
     c. A benign symptomatic lesion to be treated with elective surgery and no further work-up.
     d. A lesion with uncertain diagnosis regarding benign or malignant status, requiring biopsy; it is the radiologist's role to keep this group as small as possible.

      e. A malignant lesion that needs a preoperative radiologic and clinical work-up, followed by surgery for confirmation of histology, and, in many cases, definitive treatment; if so, the work-up should be tailored to the individual patient, depending on the set of most likely diagnoses and the radiologist's responsible analysis of costs and benefits of various examinations.

2. Guidance of surgical resection: The radiologist must understand the natural history of the lesion and must be conversant with the surgeon's treatment options, in order to give a complete assessment of tumor involvement; the radiologist must also be aware of which diagnostic modality is most suitable for this purpose for each tumor.

3. Knowledgeable post-treatment follow-up: Given the natural history of the lesion and the treatment in the individual case, the radiologist should know which diagnostic modalities can most effectively monitor for recurrence and complications.

B. The description of a tumor on a radiograph *must* include an assessment of the following *ten determinants*. If they are accurately assessed, the diagnosis (or two or three most likely diagnoses) usually becomes obvious. Generalizations applicable to each determinant are discussed below.

1. Age of patient: This is a more important determinant in some lesions than others and may occasionally lead you to the correct diagnosis when the tumor is otherwise atypical; common tumors in various age groups are as follows:

Age (yr)	Lesion
1	Metastatic neuroblastoma
1–10	Ewing's sarcoma (tubular bones)
10–20	Aneurysmal bone cyst
10–30	Osteosarcoma, Ewing's sarcoma (flat bones)
Skeletally immature	Chondroblastoma
Skeletally mature–50	Giant cell tumor (GCT)
30–60	Chondrosarcoma, primary lymphoma, Malignant fibrous histiocytoma, Fibrosarcoma
50–80	Metastasis, multiple myeloma

2. Soft tissue involvement.
    a. Cortical breakthrough of a bone lesion to create a soft tissue mass generally suggests an aggressive lesion.
    b. Such a soft tissue mass generally distorts muscle planes but otherwise leaves them intact; infection, on the other hand, often obliterates these planes.
    c. Soft tissue sarcomas (such as liposarcoma, malignant fibrous histiocytoma [MFH], synovial sarcoma) often appear radiographically and at surgery to be very distinct and even encapsulated, leading to the mistaken assumption that they can be "shelled out" successfully; benign soft tissue tumors (especially desmoids) may, on the other hand, appear very infiltrative and locally aggressive.
3. Pattern of bone destruction.
    a. Geographic: least aggressive, well-defined margin.

   b. Moth-eaten: more aggressive; margin less well-defined.
   c. Permeative: highly aggressive, poorly demarcated, and often even very difficult to visualize.
4. Size of lesion.
5. Location of lesion.
   a. The *particular bone* may be important. For example, certain tumors—adamantinoma, ossifying fibroma, chondromyxoid fibroma—occur much more commonly in the tibia than elsewhere. Another example might be thin tubular bones, such as the ulna and fibula, where lesions such as nonossifying fibroma [NOF] and aneurysmal bone cyst [ABC], which are usually eccentrically placed, appear centrally placed owing to the small diameter of the bone.
   b. Flat vs. tubular bones: Many lesions are found more commonly in one than the other.
   c. Appendicular vs. axial skeleton: Again, many lesions are found more commonly in one than the other.
   d. Epiphysis, metaphysis, or diaphysis: Some lesions are dependably found in one region or another; for example, chondroblastoma is the only lesion frequently found in the epiphysis, osteosarcoma is most commonly metaphyseal, and Ewing's sarcoma is usually diaphyseal.
   e. Central, eccentric, or cortical epicenter of the lesion: For example, an expanded nonaggressive lytic lesion in the metaphysis of a tubular bone is more likely a solitary bone cyst, ABC, or NOF if the epicenter is located centrally, eccentrically, or cortically, respectively.

6. Zone of transition from abnormal to normal bone.
   a. Wide: Aggressive.
   b. Narrow: Nonaggressive.
7. Margin of lesion: Sclerotic or nonsclerotic.
   a. Note that "margin" is a different entity than "zone of transition." Logically, most nonaggressive lesions have a sclerotic margin and narrow zone of transition, and most aggressive lesions do not have a sclerotic margin and have a wide zone of transition; there are, however, some lesions that typically have a narrow zone of transition but no sclerotic margin (GCT and occasionally, plasmacytoma).
   b. These characteristics are important determinants in these lesions.
8. Presence of visible tumor matrix.
   a. Aggressive bone-forming tumors produce amorphous osteoid, which may be less dense than or as dense as normal bone.
   b. Less aggressive bone-forming tumors produce better organized dense bone.
   c. Cartilage-forming tumors produce a stippled matrix with C and J shapes; the matrix is denser than normal bone.
9. Host response.
   a. Cortical thickening, expansion, or penetration.
   b. Periosteal reaction.
10. Polyostotic vs. monostotic lesion: Polyostotic lesions automatically restrict the disease process to:
    a. Benign: Fibrous dysplasia, Paget's, histiocytosis, multiple exostoses, multiple enchondromatosis.
    b. Malignant: Metastases, myeloma, or primary bone tumors with bony metastases (Ewing's sarcoma, osteosarcoma, MFH).

11. After the ten major determinants are identified, a *conclusion* should be drawn that states whether the lesion is *aggressive or nonaggressive*. Note that the conclusion should *not* be whether the lesion is malignant or benign; several aggressive lesions are, in fact, benign (e.g., histiocytosis, ABC, infection).

C. Surgical staging of solitary bone tumors and soft tissue sarcomas, Enneking methodology.[1]

1. Accepted in the orthopaedic community as a guideline to both prognosis and treatment; uses both histologic and radiographic information.

2. This staging system does not apply to metastatic lesions or round cell lesions, but does include Ewing's sarcoma.

   a. G: grade (histologic).
      (1) G_0—benign.
      (2) G_1—low-grade, malignant.
      (3) G_2—high-grade, malignant.

   b. S: site (radiographic and clinical features).
      (1) T_0—true capsule surrounds lesion (reactive rim of tissue).
      (2) T_1—extracapsular, intracompartmental compartments are defined as follows:
         a Skin—subcutaneous.
         b Parosseous—a potential compartment is seen when a lesion pushes muscle away from bone without invading either muscle or cortex.
         c Bone—intracortical; also a lesion in ray of the hand or foot is considered intracompartmental.
         d Muscle compartments—may contain more than one muscle if the muscle group is limited by a fascial plane:
            i Posterior compartment calf.
            ii Anterior compartment calf.
            iii Anterolateral compartment calf.

    iv Anterior thigh.
     v Medial thigh.
    vi Posterior thigh.
    vii Buttocks.
    viii Volar forearm.
    ix Dorsal forearm.
     x Anterior arm.
    xi Posterior arm.
    xii Deltoid.
    xiii Periscapula.

  (3) T_2—extracapsular, extracompartmental extension from any of the above-named compartments or abutment of major neurovascular structures; in addition, some sites are extracompartmental by origin:
    a Midhand, dorsal or palmar.
    b Mid- or hindfoot.
    c Popliteal fossa.
    d Femoral triangle.
    e Obturator foramen.
    f Sciatic notch.
    g Antecubital fossa.
    h Axilla.
    i Periclavicular.
    j Paraspinal.
    k Periarticular, elbow or knee.

 c. M: metastases.
  (1) M_0—no metastases.
  (2) M_1—metastases present.
 d. Staging.
  (1) *Benign:*

1 (Inactive)	2 (Active)	3 (Aggressive)
G_0	G_0	G_0
T_0	T_0	T_{1-2}
M_0	M_0	M_{0-1}

(2) *Malignant:*

Ia: Low-grade without metastases, intracompartmental.
Ib: Low-grade without metastases, extracompartmental.
IIa: High-grade without metastases, intracompartmental.
IIb: High-grade without metastases, extracompartmental.
III: Low- or high-grade with metastases.

Ia	Ib	IIa	IIb	III
G_1	G_1	G_2	G_2	G_2
T_1	T_2	T_1	T_2	T_2
M_0	M_0	M_0	M_0	M_1

D. Surgical treatment options.
 1. Intralesional excision (curettage): Tumor is found at the margin.
 2. Marginal excision (excisional biopsy): Plane of dissection passes through the reactive tissue or pseudocapsule of the lesion; satellites of residual lesion may be found; inadequate for malignant or recurrent benign lesions.
 3. Wide excision: Removal of lesion surrounded by an intact cuff of normal tissue. The plane of dissection is well beyond the reactive tissue surrounding the lesion, but the entire muscle or bone is not removed. Because the plane of resection passes through the compartment skip lesions may remain. Adequate for recurrent, aggressive benign tumors, low-grade sarcomas, and some high-grade sarcomas that have been reduced in bulk by chemotherapy.

4. Radical resection: Removal of the lesion along with the entire muscle, bone, or other involved tissues in the compartment (bones removed joint to joint and muscles removed origin to insertion).

5. Any of these margins can be achieved by either amputation or limb salvage procedures. Limb salvage *en bloc* procedures most commonly fall in the category of wide excisions, offering tumor control without sacrifice of limb. Consideration of limb salvage is based on the staging of the lesion, anatomical location, age (and expected growth) of the patient, extent of local disease, expected function after the procedure, and the efficacy of adjuvant therapy. Studies have shown that the significant factor in tumor surgery is the margin achieved rather than the method of achieving that margin (limb salvage vs. amputation).

E. Philosophy of tumor work-up.

Once a lesion is deemed aggressive and biopsy is planned, the diagnostic work-up should be tailored to the individual lesion to (1) stage the lesion and (2) assist in treatment planning. Preoperative staging should have the following goals: (1) avoid overly extensive (and overly expensive) staging; (2) avoid overly aggressive surgery on low-grade lesions; (3) avoid irreversible undertreatment of aggressive lesions; and (4) if the aggressiveness of the lesion cannot be determined by plain film, it should be completely worked up to determine stage, assuming aggressiveness. Staging is done radiographically by discovering the presence or absence of metastatic disease and by determining whether the lesion is encapsulated, intracompartmental, or extracompartmental. For treatment planning, communication with the surgeon is essential to ensure that the correct questions are addressed and answered by the diagnostic procedures. For example, if amputation is the only surgical option, the work-up need be directed only toward defining the proximal extent of the lesion. If, on the other hand, limb salvage is considered, both the proximal and distal extent of bone and muscle involvement must be determined, as well as involvement of the vital neurovascular structures.

Choice of biopsy site should also be guided by the radiographic work-up. Biopsy must be made of the most aggressive viable portion of the lesion. Typically, the central portion of an aggressive lesion is necrotic, so more information is gained from a peripheral biopsy. Biopsy should be made through an approach that can be resected at the definitive surgical procedure and that, therefore, does not (through possible tumor spill) compromise vital tissue planes and does not compromise skin that may be needed to close over a resected area. The biopsy site should be within a single compartment and should not approach neurovascular structures.

Percutaneous biopsy by the radiologist has become quite popular since it may reduce both cost and morbidity. The site must be chosen carefully, as described above, and at least three needle passes are made. It should be remembered that histologic diagnosis of metastatic disease and myeloma can be made with very little tissue but that diagnosis of primary malignant bone tumors requires larger amounts of tissue. Primary benign bone tumors often require tissue obtained by open biopsy, since larger specimens yet are needed for histologic diagnosis.

Bearing in mind the need to arrive at a logical differential diagnosis, the need to stage the lesion correctly, and the need to assist in biopsy and treatment planning, the radiographic work-up should proceed as follows:

1. Plain film.
   a. Best method for assessing bone detail, aggressiveness of the lesion, and certain diagnostic features, as detailed in Section B of this chapter.
   b. Extent of lesion often can be determined.
   c. Used to assess whether further work-up is required.
2. Radionuclide studies.
   a. Technetium ^{99m}Tc MDP scans are used primarily to determine whether a lesion is monostotic or polyostotic. Such a study is therefore essential in staging a bone tumor. Although the degree of abnormal uptake may be related to the aggressiveness of the lesion, this does not correlate with histologic grade, and, in fact, some benign lesions, such as osteoid osteoma, show very significant uptake owing to hypervascularity and host bone reaction. A bone scan may not accurately demonstrate extent of lesion: some lesions show an "extended uptake" pattern beyond the margin of tumor, perhaps secondary to hyperemia.
   b. Gallium ^{67}Ga may show uptake in a soft tissue sarcoma and may help differentiate a sarcoma from a benign soft tissue lesion (sensitivity 85%, specificity 92%).[2]
3. Computed tomography (CT).
   a. May add diagnostic features to the lesion (for example, matrix calcification may be better seen).
   b. Complements the plain film in defining degree of aggressiveness of the lesion.
   c. By measuring Hounsfield units within the marrow, osseous extent can be defined very clearly. Accuracy is much better in

diaphysis than metaphysis or epiphysis.

d. Extraosseous relations are much better defined than by plain film; thus CT may be essential in staging, and in planning biopsy and surgical treatment.

e. Remember always to image the contralateral extremity so that intra- and extraosseous extent can better be identified.

f. Disadvantages: Restricted to axial views or reconstructions in other planes; soft tissue contrast may be limited, especially in a thin person; large dose of intravenous (IV) contrast medium may be required.

4. Magnetic resonance imaging (MRI).

a. Superior to CT in defining extraosseous relations since soft tissue contrast is superb; T1-weighted images enhance contrast with fat, while T2-weighted images enhance contrast with muscle: neurovascular structures are seen well without the use of contrast medium. It is very difficult (as with CT) to differentiate tumor from peritumoral edema, so soft tissue involvement often is overestimated.

b. Thought to be as accurate as CT in defining marrow extent of the lesion; T1-weighted images are required for this, since it enhances the contrast between fatty marrow and the lesion.

c. Low specificity at present time.

d. Direct coronal, sagittal, and axial planes are all available.

e. Disadvantage: Neither calcification nor cortex is seen, though alterations in the cortex usually can be inferred.

f. Tumor work-ups usually require either CT or MRI but rarely should require both because of high cost relative to benefit. Therefore, the more appropriate examination should be chosen at the outset, depending on the questions that need to be answered: An entirely intraosseous lesion may be examined best by CT, while an entirely extraosseous lesion is far more clearly delineated by MRI. An intraosseous lesion that extends into the soft tissues may be examined by either modality, but many authors feel that MRI more easily answers the pertinent staging and surgical treatment questions.[3]

   g. Two sequences are usually obtained, in order to best contrast the lesion with surrounding tissue, whether they are fat or muscle; a lesion may appear isointense with surrounding soft tissues on any single sequence.

   h. Different pulse sequences have not, at this writing, achieved the ability to make specific histologic diagnoses based on tumor signal intensity.

5. Tomography: used only occasionally, when more detailed views of bony architecture are required.

6. Angiography: Rarely used to diagnose or define the tumor, but may be used if embolization of a highly vascular lesion is required prior to surgery. The study should be tailored to the individual patient; for example, a popliteal fossa angiogram should be performed in the lateral position, once with the knee flexed and once with the knee extended and using vasodilators in each injection to determine vascular involvement by tumor.

7. Chest film and CT for evaluation for metastatic disease.

F. Complications of tumor therapy:

  1. Radiation therapy:

  a. Tumor recurrence.
  b. Growth cessation and/or deformities if epiphyses or apophyses are included in the field of radiation.
  c. Radiation-induced osteochondroma.
  d. Increased susceptibility to infection.
  e. Radiation osteonecrosis: A permeative change seen in bones restricted to the radiation port, often appearing aggressive and producing pathologic fracture. This may be difficult to differentiate from tumor recurrence radiographically, but tends to occur 7 to 10 years after therapy.
  f. Radiation-induced sarcoma: A sarcoma (osteosarcoma, MFH, fibrosarcoma, or chondrosarcoma) arising in a previously irradiated bone, usually 4 to 20 years after therapy. These sarcomas may be difficult to differentiate initially from radiation osteonecrosis, but their aggressive nature is soon demonstrated, and prognosis is very poor.
2. Chemotherapy.
  a. Tumor recurrence.
  b. Increased incidence of infection.
3. Limb salvage with allograft.
  a. Tumor recurrence.
  b. Increased incidence of infection (large nidus of dead bone, hardware, an immunocompromised patient).
  c. Hardware failure.
  d. Graft resorption.
  e. Late fracture.
  f. Delayed union: May take up to 2 years for union with the graft to be demonstrated; limb must be protected from weight bearing during this process; union with vascularized fibular grafts may progress considerably faster.

The bone tumors and tumorlike conditions will be discussed in text in the following order. Categories are indicated by roman numerals in the text as they are listed in the outline below, which represents a modification of the WHO classification.

I. Bone-forming tumors.
   A. Benign.
      1. Osteoma.
      2. Enostosis.
      3. Osteoid osteoma.
      4. Osteoblastoma.
      5. Ossifying fibroma.
   B. Malignant.
      1. Conventional osteosarcoma.
      2. Telangiectatic osteosarcoma.
      3. Parosteal osteosarcoma.
      4. Periosteal osteosarcoma.
      5. Low-grade intraosseous osteosarcoma.
      6. Osteosarcoma of the jaw.
      7. Osteosarcoma in the older age group.
      8. Multicentric osteosarcoma.
      9. Soft tissue osteosarcoma.
II. Cartilage-forming tumors.
   A. Benign.
      1. Chondroma.
      2. Osteochondroma.
      3. Chondroblastoma.
      4. Chondromyxoid fibroma.
      5. Juxtacortical chondroma.
   B. Malignant.
      1. Chondrosarcoma.
III. GCT.
IV. Marrow tumors.
      1. Ewing's sarcoma.
      2. Primary lymphoma.
      3. Hodgkin's disease.
      4. Multiple myeloma/plasmacytoma.
V. Vascular tumors.
   A. Benign.

        1. Hemangioma.
        2. Lymphangioma.
        3. Cystic angiomatosis.
        4. Massive osteolysis.
        5. Glomus tumor.
    B. Indeterminate for malignancy.
        1. Hemangiopericytoma.
        2. Hemangioendothelioma.
    C. Malignant.
        1. Angiosarcoma.
  VI. Other connective tissue tumors.
    A. Benign.
        1. Fibromatoses (soft tissue, intraosseous, and cortical desmoid).
        2. Lipoma (intraosseous, extraosseous, and parosteal).
        3. GCT of tendon sheath.
    B. Malignant.
        1. Fibrosarcoma, MFH.
        2. Liposarcoma.
        3. Synovial cell sarcoma.
  VII. Other tumors.
        1. Chordoma.
        2. Adamantinoma.
        3. Neurofibroma.
  VIII. Tumorlike lesions.
        1. Solitary bone cyst (SBC).
        2. ABC.
        3. NOF/benign fibrous cortical defect.
        4. Eosinophilic granuloma.
        5. Fibrous dysplasia.
        6. Brown tumor of hyperparathyroidism (HPTH).
        7. Myositis ossificans.
  IX. Metastases.

The organization in each section is as follows:

    A. Eleven determinants.
        1. Age.
        2. Soft tissue involvement.

        3. Pattern of bone destruction.
        4. Size of lesion.
        5. Location of lesion.
        6. Zone of transition.
        7. Margin of lesion.
        8. Tumor matrix.
        9. Host response.
        10. Polyostotic vs. monostotic.
        11. Other features.
    B. Aggressiveness of lesion.
    C. Major differential diagnoses.
    D. Metastatic potential.
    E. Radiographic work-up.
    F. Treatment.

I. BONE-FORMING TUMORS
A. Bone-Forming Tumors: Benign

Osteoma

A hamartomatous process, with abnormal proliferation of bone and no stromal abnormalities.

> **Key Concepts:** Sclerotic; calvarium or sinuses; may be associated with Gardner's syndrome.

A. Determinants:
    1. Age: Tends to be seen in adults.
    2. Soft tissue involvement: None.
    3. Pattern: Geographic; may expand adjacent bony margins.
    4. Size: Usually greater than 2 cm.
    5. Location: Membranous bone—calvarium, usually arising from the external table, or paranasal sinuses.
    6. Zone of transition: Narrow.
    7. Margin: Entire lesion is sclerotic.
    8. Tumor matrix: Dense homogeneous bone.
    9. Host response: None.
    10. May be polyostotic.

   11. Other features: May be part of the autosomal-
dominant Gardner syndrome, occurring with
multiple adenomatous colonic polyps.
   B. Nonaggressive.
   C. Major differential diagnoses:
      1. Blastic metastasis.
      2. Hyperostosis from meningioma in calvarium
(however, the latter usually involves the
inner table).
   D. Metastatic potential: None.
   E. Radiographic work-up: Plain film diagnosis.
   F. Treatment: None.

Enostosis (Bone Island)

A hamartomatous proliferation of bone.

> **Key Concepts:** Small round sclerotic lesion;
> may be polyostotic and be mistaken for sclerotic
> metastasis.

   A. Determinants:
      1. Age: Any.
      2. Soft tissue involvement: None.
      3. Pattern: Geographic, round.
      4. Size: 2 mm to 2 cm.
      5. Location: Medullary canal of any bone.
      6. Zone of transition: Narrow, blending over a
very short distance peripherally into normal
bone.
      7. Margin: Entire lesion is sclerotic.
      8. Tumor matrix: Dense homogeneous bone.
      9. Host response: None.
      10. May be polyostic (at one end of the spectrum
of sclerosing dysplasias, including
osteopoikilosis [multiple epiphyseal
enostoses]).
      11. Other features: Very common lesion, usually
discovered fortuitously.
   B. Nonaggressive.

C. Major differential diagnoses:
  1. Blastic metastasis.
  2. Dense osteoid osteoma.
D. Metastatic potential: None.
E. Radiographic work-up: None, unless painful; with pain, either of the differentials listed might be considered. Bone scan might differentiate enostosis, which shows little or no increased uptake.
F. Treatment: None.

Osteoid Osteoma

A benign entity with distinctive radiographic and clinical findings.

> **Key Concepts:** Painful; lytic lesion with or without sclerotic nidus; sclerotic host reaction which may be distant if the lesion is intracapsular; may cause a painful scoliosis; "hot" on bone scan.

A. Determinants:
  1. Age: Second and third decades.
  2. Soft tissue involvement: Only in the rare subperiosteal form (see below).
  3. Pattern: Geographic—a circumscribed nidus (with or without calcification) surrounded by a lucent, highly vascular stroma, itself surrounded by dense reactive bone.
  4. Size: Nidus less than 2 cm.
  5. Location: The most common variety is cortically based, in the tubular bones. These elicit such a densely sclerotic reaction that the nidus may actually be masked on plain film.
      Another variety is intramedullary (and often intracapsular). These most commonly are found in the proximal femur at the medial aspect of the femoral neck (Fig 1–1). These lesions elicit a sclerotic host reaction, which

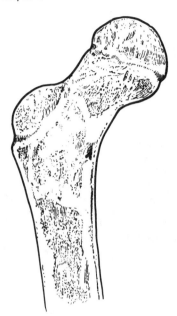

FIG 1-1.
Intracapsular osteoid osteoma with lytic nidus in the femoral neck, calcar buttressing, and mild sclerotic host reaction not immediately adjacent to the nidus. (NB: All figures are meant to correspond with radiographic appearance, with dense bone such as cortex appearing white and lytic areas appearing black.)

may be located at a considerable distance from the nidus; calcar buttressing is also a common feature. In a child, osteoid osteomas of the femoral neck may cause irreversible growth deformities (valgus, with a thick neck and overgrowth leading to limb length discrepancy), muscle atrophy, and associated early osteoarthritis.

Subperiosteal osteoid osteomas, the third, and least common, variety, demonstrate a round soft tissue mass immediately adjacent to bone with underlying scalloping and irregular bone resorption. The talus is one of the most common sites of the subperiosteal osteoid osteoma.

The spine is a relatively common site for osteoid osteoma. The posterior elements rather than the vertebral body are involved, and there is a painful scoliosis with the apex at the lesion, concave on the side of the lesion and without a rotatory component.

6. Zone of transition: Narrow.
7. Margin: Sclerotic host reaction.
8. Tumor matrix: Nidus may or may not be calcified.
9. Host response: Varying amounts of sclerosis, depending on location (see Section A-5).
10. Rarely may be polyostotic with a nidus in adjacent bones; may also have double or multiple nidi.
11. The clinical features of aching pain, worse at night, relieved with aspirin, and dramatically relieved by excision are typical; however, the intramedullary intracapsular variety may present as a less painful synovitis. Muscle atrophy of the involved limb is common. The subperiosteal osteoid osteoma not uncommonly presents clinically as arthritis. Overall incidence: 1.6% of excised primary bone tumors.

B. Nonaggressive, but recurs if nidus is excised incompletely.
C. Major differential diagnoses:
   1. Dense cortical osteoid osteoma:

   a. Brodie's abscess.
   b. Enostosis.
   c. Subacute stress fracture.
2. Intramedullary intracapsular: If the nidus is not obvious and it is confusing radiographically, may consider a synovitis or dysplasia.
3. Subperiosteal:
   a. Arthritis.
   b. Juxtacortical chondroma.
4. Posterior elements of spine:
   a. Metastasis.
   b. Spondylolysis with sclerosis of contralateral posterior elements.
D. Metastatic potential: None.
E. Radiographic work-up: If the lesion is radiographically confusing, a bone scan is very useful. The lesion almost invariably shows significantly increased uptake, occasionally with a double density. With localization guided by the bone scan, the nidus may then be demonstrated by CT, which provides information on exact location and whether there is more than one nidus.
F. Treatment: Complete marginal excision of the nidus. Recurrence is due to incomplete excision. If there is any question regarding completeness of excision, filming of the specimen may be useful. Intraoperative bone scanning has also been described but is cumbersome. Spontaneous healing over several years has been reported.

Osteoblastoma

A rare (0.5% of bone tumor biopsies) benign tumor that is difficult to differentiate histologically from osteoid osteoma. Radiographically, it is quite distinct from osteoid osteoma. Historically, the two entities have been arbitrarily distinguished by size (osteoid osteoma, less than 2 cm; osteoblastoma, greater than 2 cm).

> **Key Concepts:** Expansile, usually nonaggressive; lucent or sclerotic; posterior elements of spine.

A. Determinants:
   1. Age: Second and third decade.
   2. Soft tissue involvement: None.
   3. Pattern: Geographic—expands bone in a fusiform shape.
   4. Size: Greater than 2 cm.
   5. Location: Most common in the posterior elements of the spine (42%; Fig 1–2); in the long bones, either metaphyseal or diaphyseal and central to eccentric.
   6. Zone of transition: Narrow.
   7. Margin of lesion: Thin rim of sclerosis.

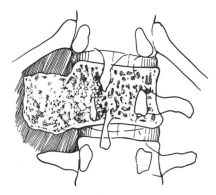

FIG 1–2.
Osteoblastoma, a geographic expanded lesion most commonly located in the posterior elements of the spine.

8. Tumor matrix: May be lucent, mixed, or completely blastic.
9. Host response: Expansion without large amount of sclerosis or periosteal reaction.
10. Monostotic.
11. Other features: Patient presents with a dull, aching pain, less severe than that of osteoid osteoma.

B. Generally, as described, appears nonaggressive; occasionally, though, the cortex may be broken through and the lesion appears more aggressive.

C. Major differential diagnoses:
1. Osteosarcoma: If the lesion appears aggressive and contains calcified osteoid.
2. Osteoid osteoma: If size is borderline.
3. Brown tumor of HPTH: Usually has other findings of HPTH.
4. ABC: Usually located more eccentrically but may be associated with osteoblastoma, especially in the spine.
5. GCT: Usually located more eccentrically, subarticular, and less well marginated.

D. Metastatic potential:
1. Generally not malignant, but there have been a few reports of malignant transformation; some of these may be a true transformation, while others were misdiagnosed osteosarcomas.
2. There are also reported cases that appear benign radiographically but are "pseudomalignant" histologically; these patients have a benign course.

E. Radiographic work-up: These usually fall into the "radiographically benign, symptomatic" category, which can go to biopsy and elective surgery without further work-up. If there is a question of aggressiveness, bone scan followed by CT will be useful to differentiate osteoblastoma from osteosarcoma.

F. Treatment: Curettage (marginal or intracapsular excision) with bone graft; recurrence is rare; if surgically inaccessible, radiation has been used.

Ossifying Fibroma

An extremely rare benign lesion found almost exclusively in the tibia and facial bones.

> **Key Concepts:** Rare; looks like NOF since it is cortically based and expansile; location in tibia is an important factor in suggesting the diagnosis.

A. Determinants.
1. Age: Second and third decades.
2. Soft tissue involvement: None.
3. Pattern of bone destruction: Geographic oval lesion.
4. Size of lesion: Varies widely.
5. Location: In the proximal third of the tibia, in the anterior cortex, often causing an anterior cortex bowing (Fig 1–3).
6. Zone of transition: Narrow.
7. Margin of lesion: Sclerotic rim.
8. Tumor matrix: May be lucent or contain osteoid and look like ground glass.
9. Host response: None except sclerotic margin.
10. Monostotic.
11. Other features: This lesion can be similar, both histologically and radiographically, to either fibrous dysplasia or adamantinoma. Since there are no other distinguishing features, the radiologist should consider this lesion whenever the *location* is appropriate (i.e., upper third of tibia in anterior cortex).
B. Initially not aggressive but may become highly aggressive with recurrence.
C. Major differential diagnoses:
1. Adamantinoma (generally more distal in the tibia).

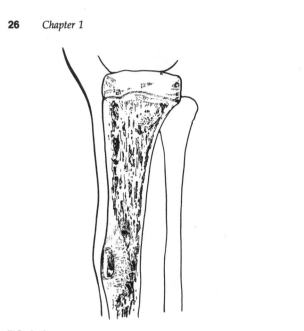

FIG 1-3.
Typical ossifying fibroma, cortically based on the anterior aspect of the proximal one third of the tibia; the same appearance at a different site would be more suggestive of NOF.

    2. Cortically based fibrous dysplasia.
    3. NOF.
D. Metastatic potential: None.
E. Radiographic work-up: Diagnosis, biopsy, and treatment are usually based on the plain film alone.
F. Treatment: Wide excision *(en bloc)* of the bony lesion; if the lesion is misdiagnosed as fibrous dysplasia or NOF and treated with curettage

(marginal excision) rather than wide excision, the recurrence rate is unacceptably high. With recurrence, the lesion behaves much more aggressively; accurate diagnosis is therefore very important.

B. Bone-Forming Tumors: Malignant

Osteosarcoma

Common (20% of primary bone tumor biopsies) primary bone tumor that produces malignant osteoid in either large or small foci. Several variants are described.

Key Concepts: Conventional osteosarcoma is extremely aggressive, with tumor matrix in soft tissue mass; childhood and adolescent tumor; metaphyseal.

Conventional Osteosarcoma.—The most common osteosarcoma (75%; Fig 1–4).

A. Determinants:
  1. Age: 10 to 25 years is most common; a smaller peak is seen in older adults (see below).
  2. Soft tissue involvement: Cortical breakthrough with large mass, often containing tumor bone.
  3. Pattern: Permeative.
  4. Size: Rarely discovered before it is large.
  5. Location: Metaphysis, 90%, diaphysis, 10%; distal femur more common than proximal tibia, which is, in turn, more common than proximal humerus.
  6. Zone of transition: Wide.
  7. Margin: No sclerotic margin.
  8. Tumor matrix: 90% produce tumor bone matrix visible on plain films or CT, but the amount varies, producing an appearance ranging from densely blastic to nearly

completely lytic. Histologically, 50% produce
enough osteoid to be termed "osteoblastic";
25% produce predominantly cartilage, and
25% produce predominantly spindle cells.
The radiographic appearance often
corresponds to these histologic findings, with
the matrix calcifications in the cartilage and
spindle cell varieties either very subtle or
entirely lacking.

It should also be noted that many lesions

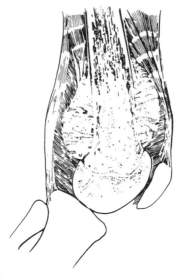

FIG 1–4.
Conventional osteosarcoma, with permeative change,
large soft tissue mass, amorphous calcification, and periosteal reaction. Metaphyseal site around the knee is
typical.

excite reactive host bone formation, which
can appear quite blastic; thus, it may be
difficult to differentiate a Ewing's sarcoma
with extensive reactive bone formation
roentgenographically from an osteosarcoma
unless amorphous tumor bone formation is
seen within the soft tissue mass (this
confirms osteosarcoma).

9. Host response: Periosteal reaction, often in an
   aggressive sunburst or Codman's triangle
   pattern.
10. Usually monostotic; however, as many as
    10% may have skip lesions of medullary
    involvement within the bone of primary
    occurrence. Most of these are too small to
    detect by plain film or bone scan.

B. Highly aggressive, radiographically as well as
   clinically (50% to 60% survival).
C. Major differential diagnoses:
   1. Ewing's sarcoma, especially if it is diaphyseal
      and has no calcified matrix in the soft tissue
      mass.
   2. Cortical desmoid: An avulsive irregularity
      posterior to the adductor tubercle. Location is
      key to the diagnosis: the cortical desmoid is
      always on the posterior aspect of the medial
      femoral condyle; it appears as a scalloped
      defect in the cortex, sometimes with a small
      soft tissue mass with periosteal elevation, and
      occasionally with matrix calcification due to
      repair. Biopsy material taken during the
      active repair phase is difficult to distinguish
      from osteosarcoma, so the lesion is best
      diagnosed by the radiologist.
   3. Early myositis ossificans: In the first 4 to 8
      weeks, amorphous calcification may occur in
      soft tissues overlying bone, and periosteal
      reaction may be seen, which is highly
      suggestive of osteo-sarcoma. Careful
      evaluation of patient history may mitigate a

potentially confusing biopsy (see Chap. 3, Myositis).
   4. Aggressive osteoblastoma: Very rare.
D. Metastatic potential: Very significant, with hematogenous spread to lungs or bone and lymphangitic spread more locally; intramedullary skip lesions may also occur. Ten percent to 20% present with metastases to lung or bone; 80% of relapses occur in lung and 20%, in bone. Those in bone do very poorly, while those in the lung have a 10% to 20% 5-year survival with resection of metastases and chemotherapy. Relapses usually occur within 2 years of the initial diagnosis.
E. Radiographic work-up:
   1. Plain film: Usually diagnostic.
   2. Bone scan: To assure it is monostotic.
   3. Chest x-ray and CT: To exclude metastatic disease.
   4. CT or MRI: To evaluate extent of lesion, plan biopsy, and plan definitive therapy.
F. Treatment (if not metastatic):
   1. Radical (amputation) or wide excision *en bloc* (limb salvage). Depending on lesion staging, a similar protocol may be undertaken in a patient with limited, resectable metastases.
   2. Chemotherapy: Still controversial; some series show no improvement in survival with chemotherapy. More recent multicenter controlled trials suggest that adjuvant polydrug chemotherapy has a favorable impact on survival for patients with nonmetastatic high-grade osteosarcoma. These studies have a relatively short follow-up period. It is possible that adjuvant chemotherapy may decrease the number of lung metastases and delay their development, thus affecting short-term survival rather than cure rate.[4, 5]

3. Induction chemotherapy prior to limb salvage surgery is viewed as very useful.
   a. Helps control the development of micrometastases in the interval needed to order the custom prosthesis or homograft.
   b. May produce regression of the primary tumor, allowing easier excision *en bloc* for limb salvage.
   c. Allows pathologic assessment of the chemotherapy regimen through evaluation of tumor necrosis at the time of limb salvage.

Telangiectatic Osteosarcoma.—A rare variant of osteosarcoma that may be difficult to diagnose radiographically, since it is entirely lytic.

A. Determinants:
   1. Age: 10 to 25 years (same as conventional osteosarcoma).
   2. May or may not have soft tissue involvement.
   3. Pattern: Geographic, though the borders may be somewhat indistinct; definitely has more of a round, blow-out appearance than the permeative, conventional osteosarcomas.
   4. Size: Large (usually greater than 5 cm).
   5. Location: Metaphyses of long bones (like conventional osteosarcoma).
   6. Zone of transition: Somewhat indistinct but much narrower than conventional osteosarcoma.
   7. Margin: Generally no sclerosis.
   8. Tumor matrix: None.
   9. Host response: May or may not have periosteal reaction.
   10. Monostotic.
   11. Lesion is highly vascular and may contain grossly visible necrotic tissue with large pools of blood and tumor at the periphery only.
B. From the description above, this lytic lesion may be confusing radiographically, having some

features that appear aggressive and others that appear much less aggressive. The radiologist must keep this lesion in mind when a moderately aggressive lytic metaphyseal lesion occurs in a teenager or young adult, since the lesion itself lacks the cardinal features of osteosarcoma (highly aggressive, with calcified matrix). Telangiectatic osteosarcoma is much more aggressive clinically than the lesions considered in the differential diagnosis and, therefore, must be treated differently.

C. Major differential diagnoses:
  1. Less aggressive looking:
     a. ABC.
     b. GCT.
  2. More aggressive looking:
     a. Fibrosarcoma/MFH.
     b. Ewing's sarcoma.
D. Metastatic potential: Great, in same distribution as conventional osteosarcoma (lungs, bones, local lymph nodes) and probably has a higher malignant potential.
E. Radiographic work-up:
  1. Plain film; considering the possibility of this the diagnosis is the key here, avoiding a misdiagnosis as a less aggressive lesion.
  2. Bone scan.
  3. Chest x-ray and CT.
  4. CT or MRI of lesion.
F. Treatment:
  1. Radical or wide marginal excision *en bloc*.
  2. Chemotherapy.

Parosteal Osteosarcoma.—Well-differentiated osteosarcoma with epicenter adjacent to the periosteum and significantly better prognosis than conventional osteosarcoma.

A. Determinants:
  1. Age: Wide range, including childhood, but 80% occur between 20 and 50 years of age

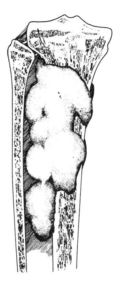

FIG 1–5.
Parosteal osteosarcoma, located typically around the knee and with densely sclerotic mature tumor matrix wrapping around the underlying bone.

       (therefore usually older than conventional osteosarcoma).

2. Soft tissue involvement: Although attached to underlying cortex at the site of origin, the lesion is otherwise located nearly entirely in the soft tissues, wrapping around the underlying bone in a lobulated fashion (Fig 1–5).
3. Pattern: Geographic.
4. Size: Often large (greater than 5 cm) at time of discovery.
5. Location: Juxtacortical, metaphyseal; 60%

distal femur; others, proximal tibia, proximal
humerus, and other metaphyseal regions.

6. Zone of transition: Usually narrow; in 10%,
   tumor extends from the cortex into the
   adjacent marrow (best evaluated by CT).
7. Margin: Entire lesion tends to be sclerotic.
8. Tumor matrix: Densely sclerotic, lobulated
   masses centrally; more peripherally these may
   be less mature bone or even a nonossified
   soft tissue mass. This zoning pattern
   distinguishes the lesion from myositis
   ossificans, in which the more mature bone is
   found peripherally.
9. Host response: None.
10. Monostotic.

B. Aggressiveness: Tends to be very slow growing
   and low grade but with inadequate excision may
   recur in a more aggressive form; with multiple
   recurrences, may dedifferentiate.

C. Differential diagnoses:
   1. Myositis ossificans: Zoning of the mature
      bone differentiates the two both
      radiographically and histologically (see A.8,
      above).
   2. Osteochondroma: Parosteal osteosarcoma
      usually has a distinct cleft between the
      underlying bone and the lesion, except at its
      origin; in addition, osteochondromas have a
      matrix of either mature bony trabeculae or the
      typical cartilaginous snowstorm appearance.
   3. Periosteal osteosarcoma.

D. Metastatic potential: Metastases to lung are often
   late (10 to 15 years) and occur much less
   frequently than in conventional osteosarcoma;
   80% to 90% 5-year survival.

E. Radiographic work-up:
   1. Plain film diagnosis.
   2. CT or MRI for evaluation of marrow
      involvement and soft tissue extent.
   3. Chest x-ray and CT.

F. Treatment: Ideal tumor for limb salvage techniques of wide marginal resection *en bloc*, retaining limb function but avoiding the recurrences possible with a marginal excision; no chemotherapy.

Periosteal Osteosarcoma.—Extremely rare surface osteosarcoma without intramedullary involvement and with a better prognosis than conventional osteosarcoma.

A. Determinants:
   1. Age: Second or third decade, usually a little later than conventional osteosarcoma.
   2. Soft tissue involvement: The mass arises juxtacortically, sometimes with perpendicular spicules of bone within the soft tissue mass.
   3. Patterns: Geographic in bone.
   4. Size: Generally 2 to 5 cm.
   5. Location: Diaphysis, most commonly of the tibia.
   6. Zone of transition: Cortex may be involved, with either thickening or "saucerization" but the zone of transition is narrow and the marrow is not involved. The extent of the entire soft tissue mass may be more difficult to determine.
   7. Margin: No sclerotic margin.
   8. Tumor matrix: Amorphous calcific densities or spicules perpendicular to the cortex may be seen in the soft tissue mass.
   9. Host response: Periosteal reaction, often with Codman's triangles.
   10. Monostotic.
B. Mildly aggressive-looking.
C. Major differential diagnoses:
   1. Conventional osteosarcoma: If any intramedullary involvement.
   2. High-grade surface osteosarcoma: Indistinguishable, though it may have a larger soft-tissue mass; this has the same histologic and survival characteristics of conventional osteosarcoma.

      3. Juxtacortical chondroma.
      4. Apophyseal avulsion with early repair.
      5. Parosteal osteosarcoma.
  D. Metastatic potential: May metastasize to lungs, but much less frequently than conventional osteosarcoma; 80% 5-year survival.
  E. Radiographic work-up:
      1. Plain film suggests diagnosis.
      2. CT or MRI defines extent (especially with regard to intramedullary involvement, which would change the diagnosis to conventional osteosarcoma).
      3. Chest x-ray and CT.
  F. Treatment: Wide marginal resection, limb salvage if possible.

Low-Grade Intraosseous Osteosarcoma.—Very rare (1% of osteosarcomas) variant that is entirely intraosseous and may be so well differentiated as to be mistaken for a benign process; excellent chance for survival if recognized initially (Fig 1–6).

  A. Determinants:
      1. Age: 10 to 25 years.
      2. Soft tissue involvement: None.
      3. Pattern: Permeative or moth-eaten.
      4. Size: Greater than 5 cm.
      5. Location: Metadiaphyseal region of long bones.
      6. Zone of transition: Wide.
      7. Margin: No sclerotic margin.
      8. Tumor matrix: Lesions range from lytic permeative to mixed sclerotic and lytic.
      9. Host response: Initially contained within bone, so no periosteal reaction; with recurrence, is more aggressive, penetrates cortex, and elicits periosteal reaction.
     10. Monostotic.
  B. Appearance ranges from mildly aggressive to more markedly aggressive.
  C. Differential diagnosis:

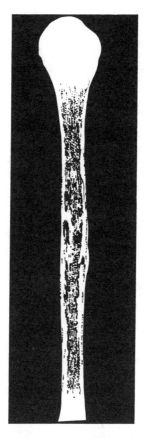

FIG 1–6.
Low-grade intraosseous osteosarcoma: permeative
diaphyseal lesion with appearance similar to Ewing's
sarcoma but no cortical breakthrough or soft tissue mass.

1. Fibrous dysplasia (if only slightly aggressive-looking).
2. Infection.
3. Ewing's sarcoma (if more aggressive-looking).
4. Lymphoma (if more aggressive-looking).
5. Fibrosarcoma/MFH (if more aggressive-looking).

D. Metastatic potential: If recognized initially and completely resected, 80% to 90% 5-year survival. With recurrence it often is highly malignant and proceeds rapidly to metastatic involvement.

E. Radiographic work-up.
   1. Plain film: May be confusing initially, having a moderately aggressive bony destructive pattern but no soft tissue mass.
   2. CT: To assess extent of marrow involvement and confirm that soft tissues are normal.
   3. Chest x-ray and CT.

F. Treatment: Initially, ideal for limb salvage with wide marginal excision *en bloc*. An aggressive recurrence may require more radical surgery plus chemotherapy.

Osteosarcoma of the Jaw.—A variant that occurs later in life than conventional osteosarcoma, has an appearance of variable destruction, may or may not have a tumor matrix, and is generally a lower grade than conventional osteosarcoma, with much better survival (80%) when treated with wide excision.

Osteosarcoma in the Older Age Group (After Age 60 Years).[6]—Generalizations:

A. Location is often different than conventional osteosarcoma:
   1. Axial skeleton 27%.
   2. Craniofacial 13%.
   3. Soft tissue 11%.

B. Eighty percent present as lytic lesions without matrix; all appear aggressive.

C. Fifty-six percent arise in a preexisting lesion:
   1. Paget's disease: Probably no more than 1% of

patients with Paget's disease are at risk for developing osteosarcoma, generally in a severely affected bone.

2. Postirradiation osteosarcoma: Generally in bone receiving more than 3000 rad; location parallels that of commonly irradiated areas (shoulder for breast carcinoma, pelvis for genitourinary tumors, and the common locations of GCT—knee, shoulder, distal radius); degenerates to osteosarcoma (50%), fibrosarcoma, or chondrosarcoma. Interval between radiation and diagnosis ranges from 3 to 40 years, averaging 14 years.

3. Dedifferentiated chondrosarcoma: Dedifferentiation occurs in up to 10% of well-differentiated chondrosarcomas. Radiographically, there is often a sharp transition between the well-differentiated chondrosarcoma and the highly destructive dedifferentiated tumor. The dedifferentiated tumor may contain elements of fibrosarcoma, MFH, high-grade chondrosarcoma, as well as osteosarcoma.

4. Osteosarcoma extremely rarely arises from benign conditions (osteochondroma, chronic osteomyelitis, osteoblastoma, bone infarct, fibrous dysplasia).

D. Survival: 37% in osteosarcoma de novo in an older patient; 7.5% in osteosarcoma that arises in a preexisting lesion.

E. Radiographic work-up:
   1. Plain film: usually diagnostic.
   2. CT or MRI for extent.
   3. Chest film and CT.

F. Treatment: Radical excision; adjunctive chemotherapy has no impact on survival at this time.

Multicentric Osteosarcoma (Osteosarcomatosis).— Synchronous appearance of osteosarcoma at multiple sites, often bilaterally symmetric; always osteoblastic, giving the appearance of bone islands (enostoses) early but rapidly progressive; extremely rare, occurring almost exclusively in children of 6 to 9 years; early development of lung metastases and extremely poor prognosis. All the lesions are about the same size, distinguishing this entity from a primary osteosarcoma with multiple bone metastases.

Soft Tissue Osteosarcoma.—Rare extraosseous osteosarcoma, generally occurring later in life (40 to 70 years). They most commonly are found in the thigh, occasionally after radiation therapy. Zoning is reversed from that of myositis ossificans, being more organized and dense centrally with a soft tissue mass peripherally. Occasionally they have no bone matrix. Prognosis is similar to that of conventional osteosarcoma with metastatic disease to lungs and local lymph nodes.

II. CARTILAGE-FORMING TUMORS

A. Cartilage-Forming Tumors: Benign

Chondroma (Enchondroma)

Common benign cartilaginous neoplasm arising in the medullary canal of bone which is asymptomatic in the absence of pathologic fracture or malignant degeneration.

> ***Key Concepts:*** Geographic lesion, usually with cartilaginous matrix; hands, feet, or metaphyses of long bones; degeneration to chondrosarcoma extremely difficult to detect radiographically: diagnosis must rely on pain and clinical suspicion.

A. Determinants:
1. Age: Often discovered incidentally in the third or fourth decade.
2. Soft tissue involvement: None.
3. Pattern: Geographic; may expand bony margins and cause cortical thinning.

4. Size: 3 to 4 cm; smaller in hands or feet.
5. Location: Central lesion.
   a. In tubular bones of the hands or feet (50%).
   b. Distributed among metaphyseal region of long tubular bones (especially the humerus, femur, and tibia), the pelvis, and the shoulder girdle (50%).
6. Zone of transition: Usually narrow, though the lesion may be lobulated.
7. Margin: Fine sclerotic margin in hands and feet; often no sclerotic margin at other sites.
8. Tumor matrix: Ranges from lytic (especially in hands or feet) to typical cartilaginous matrix (stippled, often ringlike and denser than normal bone).
9. Host response: None unless there is a pathologic fracture.
10. Usually monostotic: There may be several chondromas in the hands. If more than one chondroma is found elsewhere, the patient may have multiple enchondromatosis (Ollier's disease).

B. Not aggressive in the absence of malignant change.
C. Major differential diagnoses:
   1. In hands or feet: (NB:Chondromas are much more common than differentials c, d, e, or f, below.)
      a. Inclusion cyst (usually involves the tuft).
      b. GCT of the tendon sheath (usually causes scalloping with the epicenter in the soft tissues).
      c. GCT.
      d. ABC.
      e. Solitary bone cyst.
      f. Fibrous dysplasia.
   2. In sites other than hands or feet:
      a. Bone infarct: The pattern of calcification and absence of sclerotic margin may make

this indistinguishable from a chondroma.
b. Chondromyxoid fibroma: A rare lesion in which a calcific matrix is unusual.

D. Metastatic potential:
1. In hands or feet, the histology may appear ominous but chondroma almost universally behaves in a benign manner.
2. In sites other than hands or feet, malignant degeneration to chondrosarcoma is not uncommon. Proximal lesions (pelvis or shoulder girdle) have the highest incidence of malignant change. The difficulty lies in the fact that early malignant degeneration may not be detectable radiographically. Even serial plain films, bone scans, and CT scans may show no interval change early in the degeneration. The diagnosis, therefore, often is based solely on local pain, and appropriate treatment should be instituted on a painful chondroma in the absence of pathologic fracture. Histology in early malignant degeneration most commonly shows either low-grade chondrosarcoma or atypical chondroma.

E. Radiographic work-up:
1. If asymptomatic, plain film diagnosis.
2. If the lesion is painful and malignant degeneration is clinically suspected, CT is performed to determine extent, aggressiveness, and therapy. Note that CT appearance may be misleading: early sarcomatous change may not be manifest. Serial radionuclide bone scans may, but do not necessarily, show a change with malignant degeneration. A single bone scan is not helpful since chondromas and chondrosarcomas are both highly variable in degree of radionuclide uptake.

F. Treatment:
1. If asymptomatic (found incidentally), may

elect watchful waiting.

2. If the lesion is painful and suggests early malignant degeneration, the surgeon faces the choice of curettage and bone packing (which usually succeeds without recurrence) or complete excision *en bloc* (limb salvage, a more assuredly curative resection which carries greater morbidity). This decision is made on the basis of the patient's age and condition; thus, an elderly patient might do better with curettage since recurrence or metastatic disease is unlikely to occur during the remainder of his or her life, while a younger patient might benefit more and recover better from the curative limb salvage procedure.

G. Multiple enchondromatosis (Ollier's disease).
   a. A rare developmental abnormality characterized by the presence of enchondromas in the metaphyses and diaphyses of multiple bones.
   b. Not hereditary or familial.
   c. Appears in early childhood.
   d. Tends to be unilateral and localized to an extremity.
   e. The lesions may look like typical enchondromas or may be much larger (even grotesque in a finger).
   f. The metaphyses often have a striated appearance, with vertical lucencies and densities; most have some calcific matrix.
   g. Causes significant limb shortening and growth deformities.
   h. Risk of malignant transformation (usually to chondrosarcoma) 25% to 30%.

H. Maffucci's syndrome:
   a. Enchondromatosis, combined with soft tissue hemangiomas.
   b. Phleboliths may be present, which make the radiographic diagnosis.

c. Much higher malignant potential than enchondromatosis alone (approaching 100% according to a recent study[7]).

Osteochondroma (Exostosis)

Very common benign neoplasm that results from growth plate cartilage displaced to the metaphyseal region. Underlying bone is completely normal, and normal bone grows as an excrescence from the underlying metaphysis, with continuity of the periosteum, cortices, and marrow of the exostosis and host bone. Osteochondromas may extend from the host bone on a stalk with a cauliflowerlike head or may be much more broad based and sessile.

> **Key Concepts:** Metaphyseal; may cause growth deformity (especially if multiple and sessile) or mechanical problems (if cauliflowerlike). Growth ceases with skeletal maturity and growth or pain after skeletal maturity is suggestive of degeneration to chondrosarcoma. Multiple exostosis is hereditary (autosomal dominant) and has a 15% incidence of malignant degeneration.

A. Determinants:
1. Age: Mass is usually discovered in first or second decade.
2. Soft tissue involvement: Soft tissues are displaced by the bony mass, and a bursa may form over the mass if there is mechanical irritation.
3. Pattern: Geographic.
4. Size: Range from small to very large (10 cm).
5. Locations: Metaphyseal, with exostosis pointing away from adjacent joints; 36% are found around the knee; 95%, in the extremities.
6. Zone of transition: None, since normal bone is found within the exostosis.

7. Margins: No sclerotic margins, since normal bone extends from host bone to the exostotic bone.
8. Tumor matrix: Normal bone is within the stalk, but cartilaginous matrix may be seen in the cartilaginous cap.
9. Host response: None.
10. Ninety percent are solitary (for discussion of multiple exostoses, see Section G).
11. Other features: Osteochondromas are distinctive in that growth from the cartilage cap normally ceases with fusion of the epiphyses. Growth after maturity suggests malignant degeneration.

B. Nonaggressive in the absence of malignant change.
C. Major differential diagnoses:
    1. Stalk or cauliflower type.
        a. Parosteal osteosarcoma: Matrix is different and does not have continuity of cortex with host cortex.
        b. Chondrosarcoma: Malignant degeneration of an osteochondroma may be indistinguishable radiographically from benign osteochondroma.
        c. Myositis ossificans: No cortical continuity.
    2. Broad-based sessible type.
        a. Metaphyseal dysplasia.
        b. Fibrous dysplasia.
D. Metastatic potential: Less than 1% of solitary osteochondromas undergo malignant degeneration to chondrosarcoma. This may be detected radiographically as new mineral deposition beyond previously documented contours or, very rarely, destructive changes at the neck or base of the exostosis. More often, there is no radiographic change but the patient complains of pain or growth of the exostosis (after closure of epiphyses). In the absence of mechanical reasons for pain or the formation of a

bursa simulating growth of the exostosis such clinical symptoms indicate malignant degeneration until proven otherwise.

E. Radiographic work-up:
1. Plain film diagnosis of osteochondroma.
2. If malignant degeneration is suspected, chondrosarcoma work-up is done, which usually requires CT or MRI.
3. Bone scan shows mildly increased uptake for osteochondroma and variable uptake for chondrosarcoma; a single bone scan, therefore, often is not useful though serial scans may be.

F. Treatment.
1. Solitary osteochondroma as an incidental finding: None.
2. Osteochondroma painful for mechanical reasons: Local excision.
3. Suspected malignant degeneration: See Peripheral Chondrosarcoma, section II.B.

G. Multiple hereditary exostoses:
1. Autosomal-dominant disorder resulting in development of multiple osteochondromas.
2. Although some of these exostoses are stalklike, most are broadbased and involve a greater circumference of the metaphysis, simulating dysplasia.
3. Appears first in childhood as lumps around joints.
4. Results in shorter limbs and deformities.
5. Relatively high incidence of sarcomatous degeneration (10% to 20%), especially in more proximal lesions. Degeneration occurs earlier in life than for solitary exostosis.
6. Treatment: Resect locally as necessary for mechanical problems and observe for sarcomatous degeneration.

H. Dysplasia epiphysialis hemimelica (Trevor-Fairbank disease):

1. Intra-articular osteochondromas arising from the epiphysis.
2. May occur in single or multiple epiphyses, generally on one side of the body.
3. Knee and ankle are most common sites.
4. Histologically identical to exostosis.
5. Radiographically, a lobulated mass arising from an epiphysis, usually well mineralized.
6. Causes joint deformity, pain, and limited range of motion.
7. Treatment: Local resection.

Chondroblastoma

Fairly rare benign cartilaginous lesion found almost exclusively in the epiphysis (one of the few lesions *ever* found in epiphysis; Fig 1–7).

> **Key Concepts:** Found in epiphysis in skeletally immature patient (most commonly proximal humerus) but may extend into metaphysis with plate closure; cartilage matrix may—but need not—be present; appears benign but may elicit periosteal reactions.

A. Determinants:
1. Age: Second decade most common (before epiphyseal closure).
2. Soft tissue involvement: None.
3. Pattern: Geographic—oval and sometimes lobulated.
4. Size: 1.5 to 4 cm.
5. Location: Eccentric, with epicenter nearly always in the epiphysis; with partial epiphyseal plate closure, the lesion may involve the metaphysis; very rarely, the epicenter may be in the metaphysis or apophysis; common sites of involvement: proximal humerus, proximal femur, distal femur, proximal tibia.

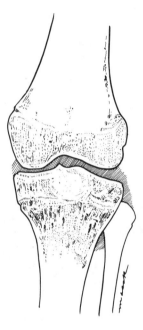

FIG 1–7.
Chondroblastoma, showing its typical nonaggressive
appearance within the epiphysis of the tibia. Although
cartilage matrix often is not appreciated on plain film,
it is more commonly seen with CT.

6. Zone of transition: Narrow.
7. Margin: Sclerotic.
8. Tumor matrix: Lytic, with cartilaginous
   calcifications in 50%.
9. Host response: Occasional periosteal reaction
   along metaphysis.
10. Monostotic.

11. Other features: Mild pain, often referred to the adjacent joint.
B. Not aggressive.
C. Major differential diagnoses:
   1. GCT crossing into the epiphysis.
   2. Articular lesions with large cysts such as pigmented villonodular synovitis (PVNS).
D. Metastatic potential: Almost negligible. There are only isolated case reports of malignant chondroblastomas with lung metastases (in the absence of radiation therapy).
E. Radiographic work-up: Plain film diagnosis.
F. Treatment: Curettage and bone graft. Recurrence rate is 15%, and recurrence is more likely if the chondroblastoma has an ABC component; recurrent lesions may be more aggressive and may require wider excision.

Chondromyxoid Fibroma

Very rare benign cartilaginous lesion also containing fibrous and myxoid tissue; found most commonly in the proximal tibial metaphysis.

> **Key Concepts:** Rare bubbly lesion found almost exclusively in proximal tibial metaphysis. That location is the major key to the diagnosis, since typical cartilaginous matrix is rare.

A. Determinants:
   1. Age: Second and third decades.
   2. Soft tissue involvement: None.
   3. Pattern: Geographic; often lobulated, with a pseudoseptate appearance.
   4. Size: Often greater than 5 cm.
   5. Location: Eccentric in metaphysis; one third in proximal tibia; others distributed in the proximal and distal femur, flat bones, and tarsals.
   6. Zone of transition: Narrow.

7. Margin: Thick sclerotic margin, with endosteal sclerosis at sites of cortical expansion.
8. Tumor matrix: Calcification rare (2%).
9. Host response: Commonly, endosteal sclerosis; periosteal reaction is rare.
10. Monostotic.

B. Not aggressive.
C. Major differential diagnoses:
1. GCT (generally does not have a sclerotic margin).
2. NOF.
3. ABC.
4. Chondroblastoma.
5. Enchondroma.

D. Metastatic potential: Malignant transformation (in the absence of radiation therapy) is extremely rare.
E. Radiographic work-up: Diagnosis is suggested by plain film and proven by subsequent biopsy; additional imaging generally is not useful.
F. Treatment:
1. Curettage and bone grafting are commonly the initial therapeutic approach; a high recurrence rate (25% to 30%) may be due to the lobulated nature of the lesion (lobules may be missed on curettage). Interestingly, younger patients tend to have more aggressive lesions and a higher recurrence rate.
2. Wide excision may be attempted after recurrence, or initially for a more aggressive lesion.

Juxtacortical (Periosteal) Chondroma

A benign cartilaginous lesion originating on the periosteal surface of the bone producing a soft tissue mass and cortical erosion that may be difficult to differentiate from more aggressive lesions.

> **Key Concepts:**   Rare lesion that appears
> aggressive, with cortical scalloping, soft tissue
> mass, and even periosteal reaction; location in
> hands or feet is suggestive.

A. Determinants:
   1. Age: Occurs in both children and adults but
      generally before 30 years of age.
   2. Soft tissue involvement: Soft tissue mass.
   3. Pattern: Geographic, with "saucerization" or
      erosion of the cortical surface.
   4. Size: Less than 4 cm.
   5. Location: Periosteal (juxtacortical) epicenter;
      most commonly in the small tubular bones of
      the hands and feet, though it may be seen in
      the larger tubular bones.
   6. Zone of transition: Narrow.
   7. Margin: Sclerotic margin is usually seen at the
      eroded cortex.
   8. Tumor matrix: Calcification in soft tissue in
      50% of lesions.
   9. Host response: Smooth periosteal reaction is
      not uncommon.
  10. Monostotic.
B. Aggressiveness: Bone involvement does not
   appear aggressive, but the soft tissue mass
   makes the lesion appear more ominous and may
   confuse the diagnosis. The histology may also
   appear more aggressive than that of central
   chondromas, but the behavior is benign.
C. Major differential diagnoses: (NB: Lesions 1
   through 3 should appear more aggressive than
   the juxtacortical chondroma and, furthermore,
   rarely appear in the hands or feet.)
   1. Chondrosarcoma.
   2. Parosteal osteosarcoma.
   3. Periosteal osteosarcoma.
   4. GCT of tendon sheath.

D. Metastatic potential: With recurrence, may undergo malignant transformation.

E. Radiographic work-up: With the typical appearance in the hands or feet, it is a plain film diagnosis. A more aggressive appearance in a large tubular bone may require CT or MRI and may present a diagnostic dilemma.

F. Treatment: Excision *en bloc* precludes recurrence and malignant transformation.

B. Cartilage-Forming Tumors: Malignant

Chondrosarcoma

Cartilage-producing sarcoma.

> **Key Concepts:** Third most common primary malignant bone tumor (following osteosarcoma and multiple myeloma); these sarcomas are often asymptomatic, leading to large tumors with late diagnosis; errors in diagnosis are also common, again leading to delay in appropriate treatment; exostotic variety almost always has cartilaginous matrix, but central variety may not; radiographically, the lesion may appear unaggressive or only mildly aggressive. If this is seen in conjunction with pain or growth of an exostosis in an adult, sarcoma should be assumed and the appropriate work-up instituted.

Central (Medullary) Chondrosarcoma.—Either primary or secondary (degeneration of enchondromas, usually those lesions located more proximally in the skeleton) chondrosarcoma arising centrally in the bone (Fig 1–8).

A. Determinants:
  1. Age: Fourth, fifth, sixth decades most common.
  2. Soft tissue involvement: Generally no soft tissue mass, although a high-grade lesion may have soft tissue involvement.

3. Pattern: Tends to be geographic, but may have portions that appear more permeative.
4. Size: Generally greater than 5 cm.
5. Location: Central, metaphyseal ends of long bones; also seen in pelvis and shoulder girdle.
6. Zone of transition: Large portions of the tumor may have a narrow zone of transition, but other regions may have a less distinct zone of transition.
7. Margin: Usually no sclerotic margin.

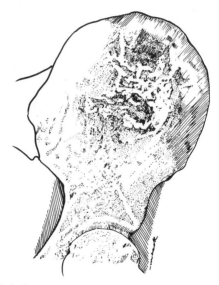

FIG 1–8.
Chondrosarcoma, commonly located in the iliac wing, demonstrating typical densely sclerotic C- and J- shaped cartilaginous matrix.

8. Tumor matrix: Central chondrosarcomas range from completely lytic to lytic with a few flecks of calcification to dense aggregates of anular calcification.
9. Host response: Periosteal reaction is variable; the endosteum may be significantly thickened (a feature that may suggest the diagnosis).
10. Monostotic.

B. Aggressiveness: 90% of central chondrosarcomas are low grade. This corresponds to the determinants listed above, which describe only a moderately aggressive lesion. The take-home lesson is this: Chondrosarcomas are common lesions and are usually not very aggressive-looking radiographically. Therefore, if a central lesion (in the right age group) appears slightly to moderately aggressive (question of permeative pattern or a questionable zone of transition), subtle calcification should be sought and the diagnosis of chondrosarcoma should be offered, whether or not cartilaginous calcification is found. This lesion is very commonly underdiagnosed since it so often appears rather benign; this underdiagnosis results in undertreatment.

C. Major differential diagnoses:
  1. If calcification is present:
     a. Enchondroma, if it looks unaggressive.
     b. Bone infarct degenerating to MFH (if it appears more aggressive).
  2. If no calcific matrix is present:
     a. MFH.
     b. Fibrosarcoma.
     c. Lymphoma.
     d. Osteosarcoma.
     e. GCT (if less aggressive).

D. Metastatic potential:
  1. Ninety percent are low-grade lesions, so local recurrence is more common than metastatic disease.

2. Metastasis involves the lungs, often quite late (even 20 years after diagnosis).
3. Five-year survival is 75%; this can be improved by:
   a. More prompt radiographic diagnosis.
   b. Meticulous biopsy technique: Chondrosarcoma is very readily implanted in soft tissues since it does not need a blood supply to survive, so recurrences may be due to tumor spill at time of biopsy or resection. Similarly, placement of an intramedullary rod may disseminate tumor cells.
E. Radiographic work-up:
   1. Plain film: Suggests diagnosis.
   2. CT: May detect subtle calcification in a lytic lesion and make the diagnosis; otherwise CT or MRI is useful to determine tumor extent and biopsy and therapy plans.
   3. Bone scan: Variable uptake is not particularly useful.
   4. Chest x-ray and CT.
F. Treatment: Radical excision or limb salvage *en bloc*. Radiation and chemotherapy are effective only as palliative therapy.

Peripheral (Exostotic) Chondrosarcoma.—Either primary or secondary (degeneration of an exostosis) chondrosarcoma arising peripheral to normal bone.

A. Determinants:
   1. Age: Third, fourth, and fifth decades.
   2. Soft tissue involvement: Unmineralized soft tissue mass may arise from the cartilaginous cap of the exostosis, usually only displacing rather than invading local soft tissue.
   3. Pattern: Geographic.
   4. Size: Greater than 5 cm, usually quite large before detected.
   5. Location: Peripheral, metaphyses of long bones as well as pelvis, shoulder girdle, sternum, and ribs.

6. Zone of transition: Generally narrow.
7. Margin: None (see section on osteochondroma).
8. Tumor matrix: Cartilage cap is often densely calcified. Peripheral streaking or a snowstorm pattern of calcification may help differentiate chondrosarcoma from exostosis. A cartilage cap more than 1 cm thick suggests malignant change to chondrosarcoma, but this may be very difficult to detect and quantify.
9. Host response: Generally none.
10. Monostotic.
11. As described in the section on osteochondroma, degeneration to chondrosarcoma often produces no clear-cut radiographic signs. Therefore clinical signs of pain and growth of exostosis after epiphyseal closure should be considered of primary importance in making the diagnosis of peripheral chondrosarcoma.

B. Generally not aggressive-looking.
C. Major differential diagnosis: Benign osteochondroma.
D. Metastatic potential: Metastases to the lungs are rather uncommon and generally late, owing to the low-grade nature of most peripheral chondrosarcomas.
E. Radiographic work-up:
1. Plain film: Watch for change in osteochondroma (growth, change in calcification pattern, destruction of osteochondroma or underlying bone), but pay strict attention to clinical signs in the absence of radiographic changes.
2. CT or MRI: CT does not reliably show the extent of the cartilaginous cap or soft tissue mass; MRI may show these more reliably.
F. Treatment: Radical or wide excision; generally no radiation or chemotherapy.

Clear Cell Chondrosarcoma.—Very rare lesion most often mistaken for a chondroblastoma because it is low grade and occurs in the ends of long bones.

A. Determinants:
    1. Age: Third decade (older than most chondroblastomas).
    2. Soft tissue involvement: None.
    3. Pattern: Geographic.
    4. Size: Less than 5 cm.
    5. Location: The ends of long bones, most commonly the femur and humerus.
    6. Zone of transition: Narrow.
    7. Margin: Sharp sclerotic margin.
    8. Tumor matrix: Generally absent.
    9. Host response: Periosteal reaction rare.
    10. Monostotic.
    11. The description is identical to that of chondroblastoma, except that it occurs in a slightly older age group, which initially may be the only hint of the true diagnosis. After growing slowly for a number of years, it may become more aggressive.

B. Generally appears nonaggressive.

C. Major differential diagnosis:
    1. Chondroblastoma.
    2. Later, if more aggressive, MFH, fibrosarcoma, or lymphoma.

D. Metastatic potential: Very low.

E. Radiographic work-up: Plain film.

F. Treatment: Wide excision; this is more aggressive treatment than that for chondroblastoma (curettage); curettage alone can result in an aggressive recurrence.

Dedifferentiated Chondrosarcoma.—The following are characteristic of this lesion:

A. Degeneration of a typically indolent chondrosarcoma into a highly aggressive lesion.

B. Degenerates to fibrosarcoma, MFH, high-grade

chondrosarcoma, osteosarcoma, or a lesion that has several elements.

C. Ten percent of chondrosarcomas may dedifferentiate. This is, therefore, not a rare lesion, since chondrosarcomas are relatively common.

D. The original chondrosarcoma and the dedifferentiated elements coexist, so the biopsy site must be chosen to include the more aggressive lesion.

E. Prognosis is poor (20% 5-year survival), and lung metastases are common.

F. Treatment is radical excision and chemotherapy.

Mesenchymal Chondrosarcoma.—The following are characteristic of this lesion:

A. Exceedingly rare, highly malignant sarcoma.

B. Age group is younger than standard chondrosarcoma (10 to 40 years).

C. Site is unusual: One third arise in soft tissues; in the skeleton, rib and jaw lesions are common while long bone lesions are unusual.

D. Calcification is usually present.

E. Radiographically aggressive; it either looks like chondrosarcoma or is nonspecific.

F. Treatment is radical resection; Hematogenous metastases may be expected to occur commonly and early.

III. GIANT CELL TUMOR (GCT)

A relatively uncommon (5% of primary bone tumors) lesion consisting of multinucleated giant cells in a fibroid stroma; these are distinct from many other lesions that may contain reactive giant cells (Fig 1–9).

> **Key Concepts:** Subarticular, eccentric lytic lesion with a narrow zone of transition but no sclerotic margin; most common around the knee or distal radius; recurrence after curettage is very common.

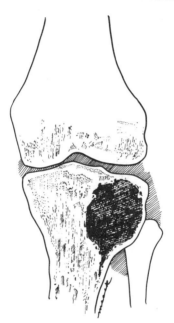

FIG 1–9.
GCT demonstrating its typically lytic nature, with a narrow zone of transition but lack of sclerotic margin. The eccentric and subarticular location is also typical.

A. Determinants:
  1. Age: Lesion nearly always occurs *after epiphyseal fusion* (85% after age 20 years); 70% fall between 20 and 40 years of age.
  2. Soft tissue involvement: Cortical breakthrough and soft tissue mass may be seen in 24%.
  3. Pattern: *Geographic expanding lesion.*
  4. Size: Often large (3 to 5 cm) by time of diagnosis.

5. Location:
    a. In tubular bones: lesion starts *eccentrically in the epiphysis;* it later *extends to the subarticular end* of the bone and may involve the metaphysis as well. A few may have a metaphyseal epicenter.
    b. *About the knee* (distal femur and proximal tibia) *or wrist* (distal radius or ulna): 65% of cases.
    c. In thin tubular bones: Appear to be central rather than eccentric.
    d. Epiphyseal equivalents may be sites of involvement in flat bones (especially around the acetabulum).
    c. In the spine: Often involves the sacrum or body of the vertebra (rarely the posterior elements).
  6. *Zone of transition: Narrow.*
  7. *Margin: No sclerotic margin.* (NB: *This combination of narrow zone of transition and no sclerotic margin is a highly reliable feature of GCT that is rarely seen in other lesions.*)
  8. Matrix: None.
  9. Host response: No periosteal or marginal reaction in the absence of fracture.
  10. Generally monostotic: May be multicentric, especially in the skull and facial bones affected by Paget's disease. In the case of multicentric GCTs, hyperparathyroidism with multiple brown tumors should be excluded.
B. Aggressiveness: Generally appear unaggressive but occasionally destroy cortex and present with a soft tissue mass.
C. Major differential diagnoses:
  1. In long bones:
    a. Chondroblastoma: generally skeletally immature with sclerotic margin.
    b. Brown tumor of HPTH: usually have subperiosteal resorption or other signs of HPTH.

      c. NOF: metaphyseal with sclerotic border.
      d. Chondrosarcoma: if epiphyseal and only
         slightly aggressive.
      e. Plasmacytoma.
    2. In flat bones or spine:
      a. Aneurysmal bone cyst.
      b. Osteoblastoma.
      c. Chordoma.
      d. Plasmacytoma.
      e. Metastasis.
D. Metastatic potential:
    1. Majority are benign: grade I or II.
    2. Malignant (often secondary to previous
       radiation) with metastases to lungs: 15%.
    3. Of those that metastasize, prognosis may be
       good (25% mortality) with surgical resection
       of lung metastases.
    4. Problem: *One cannot reliably differentiate benign
       from malignant GCT radiographically;* similarly,
       the histologic features are not predictive of
       the tumor's ultimate behavior.
E. Radiographic work-up:
    1. Plain film diagnosis: The radiographic
       appearance and location are quite
       characteristic, and the diagnosis is usually
       easily made.
    2. CT or MRI may be required, depending on
       the type of surgery anticipated.
F. Treatment:
    1. Curative treatment would be to regard the
       lesion as a low-grade malignant neoplasm
       and treat it with a wide resection *en bloc*, after
       which the rate of recurrence is only 10% to
       15%. However, the subarticular nature of the
       lesion often requires resection of the joint and
       fusion or the placement of a long-stemmed
       custom prosthesis; in a young patient such a
       prosthesis almost inevitably loosens and
       requires multiple revisions throughout life,
       causing considerable morbidity.

2. To avoid the potential morbidity of wide resection *en bloc*, curettage and bone grafting may be the initial treatment. *A recurrence rate of 55% is expected following curettage.*

3. A compromise seems to be initial treatment with curettage and cryosurgery if resection *en bloc* seems unacceptable initially. Another method is wide resection at the metaphyseal margins and curettage of the subarticular margin; recurrence rates seem to be lower with these methods.

4. Radiation therapy should be avoided: besides having an unacceptable cure rate of only 50%, 10% to 15% of irradiated GCTs develop radiation sarcomas, with extremely high mortality. In the Mayo Clinic series, nearly 100% of malignant GCTs were secondary to radiation treatment. In rare cases when a GCT is unresectable radiation may be the only treatment option.

IV. MARROW TUMORS

The "round cell" tumors; *round cell lesions include tumors (Ewing's sarcoma, lymphoma, multiple myeloma), metastatic neuroblastoma, histiocytosis, and infection. These lesions often are not easily distinguished radiographically.*

Ewing's Sarcoma

A highly malignant round cell tumor found in children, it is the *second most common primary bone tumor in children,* after osteosarcoma (Fig 1–10).

> **Key Concepts:** Highly aggressive, usually diaphyseal with large soft tissue mass and periosteal reaction. Dense host bone sclerosis is common, but no matrix is seen in the soft tissues. Most frequently affected are tubular bones in young children and flat bones in adolescents and young adults.

A. Determinants:
   1. Age: 4 to 25 years (95%), most commonly 5 to 14 years.
   2. Soft tissue involvement: A soft tissue mass is almost invariably present.
   3. Pattern: *Permeative;* rarely expands the bone.
   4. Size: Greater than 5 cm.
   5. Location: *Central* (based in medullary canal), *most commonly diaphyseal* but metaphyseal

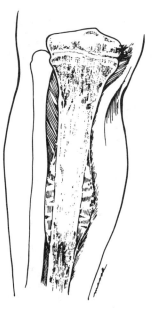

FIG 1–10.
Ewing's sarcoma: diaphyseal, permeative aggressive lesion with a large soft tissue mass. Reactive sclerosis within the bony portion of the lesion is seen.

epicenter is not uncommon; 75% involve the *pelvis or long tubular bones*. Other sites are shoulder girdle, rib, and vertebral body. Tends to involve the tubular bones in children under age 10 years and the axial skeleton, pelvis, and shoulder girdle in older children.

6. Zone of transition: Wide.
7. Margin: No sclerotic margin.
8. Matrix: *No tumor matrix is produced, but sclerotic reactive bone may be seen:* 62% are completely lytic; 23% have minimal reactive bone; 15% have marked sclerotic reactive bone. Reactive bone, however, is not found in the soft tissue mass, helping to differentiate Ewing's from an osteosarcoma with tumor bone formation in the soft tissue mass.
9. Host response:
   a. Aggressive periosteal reaction.
   b. Sclerotic reactive bone formation in medullary canal.
   c. Rarely, thick reactive endosteal bone is seen.
10. Initially monostotic, but metastases to bone are common, so the lesion may appear to be polyostotic.
11. Other features:
    a. *One third of patients with Ewing's present with fever, leukocytosis, and elevated erythrocyte sedimentation rate (ESR), simulating infection.* The overlying skin may even be warm and red. Clinical findings may, therefore, be misleading.
    b. Ewing's is extremely rare in black persons.
B. Highly aggressive appearance.
C. Major differential diagnoses:
   1. Primarily other round cell lesions: osteomyelitis, histiocytosis (may appear

highly aggressive), neuroblastoma metastasis, low-grade intramedullary osteosarcoma.

2. The duration of symptoms may be helpful in differentiating the round cell lesions. Histiocytosis (eosinophilic granuloma) may be one of the most aggressive locally, with the shortest time course (1 to 2 weeks). Osteomyelitis has a relatively short course of destruction (2 to 4 weeks), and Ewing's sarcoma has a slower course (destructive changes seen at 6 to 12 weeks).

D. Metastatic potential:
   1. Five-year survival, 50%.
   2. Fifteen to 30% have metastases at time of diagnosis; *metastases affect lung and bone with equal frequency.* Central and larger lesions have a worse prognosis than more distal ones.

E. Radiographic work-up:
   1. Plain film suggests diagnosis.
   2. Bone scan to rule out polyostotic lesion or metastases.
   3. Chest x-ray and CT.
   4. CT or MRI to determine extent and biopsy site.

F. Treatment is controversial.
   1. Combined radiation and chemotherapy: high complication rate, with sarcomatous degeneration and growth disturbance.
   2. Some protocols now call for postirradiation resection because local recurrence is found at autopsy following radiation alone in 23% of patients. Local recurrence is difficult to differentiate radiographically from radiation osteonecrosis and tumor necrosis. Since viable tumor cells are found in so many autopsy cases, wide excision following radiation and chemotherapy may be logical in selected cases; it is too early to determine the efficacy of such a protocol.

Primary Lymphoma of Bone (Reticulum Cell Sarcoma)

Uncommon lymphoma that arises initially in bone.

> **Key Concepts:** Permeative and aggressive, as other round cell lesions; soft tissue mass may be enormous.

A. Determinants.
   1. Age: Ten to 60 years, but usually 30 to 60 years.
   2. Soft tissue involvement: *Soft tissue mass may be huge* since it enlarges very rapidly and is usually asymptomatic until pathologic fracture develops.
   3. Pattern: *Permeative lytic.*
   4. Size: Very large.
   5. Location: *Central appendicular* sites are more common than axial; in descending order of occurrence: femur, pelvis, tibia, humerus, scapula.
   6. Zone of transition: Wide.
   7. Margin: No sclerotic margin.
   8. Matrix: Usually lytic (77%) but some reactive sclerosis may be seen.
   9. Host response: Often have periosteal reaction.
   10. Monostotic.
B. Aggressive appearance.
C. Major differential diagnosis:
   1. Fibrosarcoma/MFH.
   2. Osteomyelitis.
   3. Histiocytosis.
   4. Ewing's sarcoma.
D. Metastatic potential:
   1. Five-year survival 55%.
   2. Metastasizes to lymph nodes and bone; lung

metastases are unusual, but, when present, may develop and grow extremely fast (much faster than other metastatic lung nodules).
    E. Radiographic work-up:
        1. Plain film suggests diagnosis.
        2. Bone scan to determine whether monostotic.
        3. Chest x-ray and CT.
        4. CT or MR to determine local extent and lymphadenopathy.
    F. Treatment:
        1. Whole bone radiation.
        2. Chemotherapy for disseminated disease.

Hodgkin's Disease

In bone, is almost always secondary to a primary lymph node lesion; 20% of patients with Hodgkin's disease have radiographic evidence of bone involvement; extremely rare as a primary bone tumor.

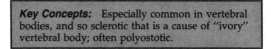

Key Concepts: Especially common in vertebral bodies, and so sclerotic that is a cause of "ivory" vertebral body; often polyostotic.

A. Determinants:
    1. Age: second through fourth decade.
    2. Soft tissue involvement: Mass is common.
    3. Pattern: Often geographic with some permeative areas.
    4. Size: Wide range.
    5. Location: *Axial skeleton* (77%) much more common than appendicular; especially common in *vertebral bodies*.
    6. Zone of transition: wide.
    7. Margin: May or may not show sclerosis.
    8. Matrix: *Sclerotic reactive bone often seen.*
        a. Lytic, 25%.
        b. Blastic, 15%.
        c. Mixed, 60%.

d. The *ivory vertebra* is a classic manifestation of Hodgkin's disease, though it is seen also in Pagets and other metastatic disease processes.
  9. Host response: Minimal periosteal reaction.
  10. Sixty-six percent polyostotic.
B. Moderately aggressive in appearance.
C. Major differential diagnosis: Metastatic disease.

Multiple Myeloma

A neoplastic proliferation of plasma cells; the most common primary bone tumor with the most common clinical presentation being back pain and anemia; both multiple myeloma and its solitary form (plasmacytoma) will be discussed.

> **Key Concepts:** Plasmacytoma may be expansile and relatively non-aggressive; multiple myeloma may present with punched-out lytic lesions or merely generalized "osteoporosis" with compression fracture. Bone scan and skeletal survey are complementary studies, since each misses a significant number of myeloma lesions.

A. Determinants:
  1. Age: Most (95%) over 40 years.
  2. Soft tissue involvement: Small soft tissue mass is common.
  3. Pattern:
    a. *Seventy percent are multifocal myeloma with focal punched-out lesions*—a moth-eaten pattern.
    b. *Fifteen percent show generalized infiltration without focal lesions:* appears only as a generalized osteopenia.
    c. *Thirty percent are plasmacytomas with an expansile geographic pattern* (which may have permeative components).

4. Size:
   a. Multiple myeloma lesions generally are less than 5 cm.
   b. Plasmacytomas usually are greater than 5 cm.
5. Location:
   a. Multiple myeloma: Originates in the red marrow but progresses to cortex and other areas. *The skull, vertebral bodies, and ribs are most commonly involved, followed by the proximal appendicular skeleton.*
   b. Plasmacytoma: Involves vertebral bodies, pelvis, femur, and humerus most commonly; rib and skull lesions are much less commonly seen.
6. Zone of transition: Relatively narrow in both the punched-out lesions of multiple myeloma and the more cystic lesions of plasmacytoma.
7. Margins: No sclerotic margin except in sclerosing myeloma (see 11, below).
8. Matrix: None except in sclerosing myeloma.
9. Host response: None.
10. Polyostotic if multiple myeloma. Plasmacytoma progresses to multifocal disease.
11. Other features:
    a. Ten to 15% are associated with amyloidosis. When amyloid is in the synovium, it simulates rheumatoid arthritis.
    b. One percent may be sclerosing myeloma, with either a sclerotic margin around the lesion or else entirely sclerotic lesions. Sclerosing myeloma is associated with a syndrome with the acronym POEMS: *p*olyneuropathy, *o*rganomegaly, *e*ndocrinopathy, *m*yeloma, *s*kin changes.
B. Appearance: Mildly aggressive.
C. Major differential diagnoses:

 1. Multiple myeloma:
    a. Metastatic disease (the latter usually involves the pedicles, not the vertebral body, as does myeloma).
    b. Osteoporosis.
    c. HPTH.
    d. Osteopoikilosis (if sclerotic).
 2. Plasmacytoma.
    a. GCT.
    b. Chondrosarcoma (intramedullary).
D. Metastatic potential:
 1. Multiple myeloma: 5-year survival 10%.
 2. Plasmacytoma: Most go on to multifocal or generalized disease, though a few remain localized; 5-year survival 30%.
E. Radiographic work-up:
 1. This is controversial, but it seems that skeletal surveys and bone scans are complementary studies in multiple myeloma.
    a. ^{99m}Tc bone scanning is positive in 25% to 40% of myeloma lesions.
    b. ^{99m}Tc is less reliable than a skeletal survey in a lesion search but is more sensitive than plain film in 18% of lesions.
    c. Plain film is more sensitive than ^{99m}Tc in 38% of lesions.
    d. MRI appears to be quite sensitive and may be helpful when one must resolve a discrepancy between scan and plain film results.
 2. Other work-up: (NB: *20% of plasmacytoma patients* will have neither serum electrophoresis or bone marrow aspirate abnormalities; in these cases, biopsy of the lesion is diagnostic.)
    a. Serum electrophoresis.
    b. Bone marrow aspiration.
F. Treatment:
 1. Multiple myeloma: Chemotherapy and

palliative radiation therapy for painful lesions and large lesions likely to develop pathologic fracture.

2. Plasmacytoma: Radiation therapy; occasionally ablative surgery is performed.

V. VASCULAR TUMORS
A. Vascular Tumors: Benign

Hemangioma

Hamartomatous lesion composed of vascular channels: *Soft tissue hemangiomas are characterized by a mass and calcified phleboliths and a characteristic can-of-worms appearance of tortuous dilated vessels on contrast enhanced CT.* Soft tissue hemangiomas may invade and erode bone, but the epicenter is clearly soft tissue. Intraosseous hemangiomas are discussed below.

> *Key Concepts:* Vertebral bodies and skull are most commonly involved and show dense vertical striations.

A. Determinants.
1. Age: fourth and fifth decades.
2. Soft tissue involvement: May have a soft tissue mass; in the vertebral bodies, this may lead to neurologic symptoms.
3. Patterns: Geographic cystic lesion.
4. Size: Usually greater than 2 cm.
5. Location: *75% in vertebral bodies (especially thoracic spine), skull (involved outer table, normal inner table), and facial bones.*
6. Zone of transition: Narrow.
7. Margin: Sclerotic.
8. Tumor matrix: None.
9. *Host response is characteristic and makes the diagnosis: In the vertebral bodies there is a coarsened vertical trabecular pattern without collapse; in the skull there are also coarsened*

trabeculae, but they radiate in a sunburst pattern to the expanded outer table.
       10. Usually monostotic.
    B. Not aggressive.
    C. Major differential diagnoses:
        1. Multiple myeloma: Usually no coarsened trabecula.
        2. Paget's disease of bone: Usually enlarged vertebral body.
        3. Metastatic disease.
    D. Metastatic potential: None.
    E. Radiographic work-up:
        1. Plain film is diagnostic.
        2. If symptomatic spine lesion, may need CT or MRI to define epidural extent.
    F. Treatment:
        1. Usually none.
        2. If symptomatic, resection and/or radiation.

Lymphangioma
  Hamartomatous lesion of dilated lymphatic vessels.

    A. Determinants:
        1. Age: Usually adult.
        2. Soft tissue involvement: None.
        3. Pattern: Geographic or moth-eaten.
        4. Size: Greater than 2 cm.
        5. Location: Central, flat, or tubular bones.
        6. Zone of transition: Narrow.
        7. Margins: Sclerotic.
        8. Tumor matrix: None (lytic).
        9. Host response: Generally none.
       10. Monostotic or polyostotic.
       11. Other features: Lymphangiomatosis is the form of lymphangioma with multiple skeletal lesions, lymphedema, and chylous pleural effusion.
    B. Unaggressive to mildly aggressive in appearance, generally nonspecific.
    C. Major differential diagnoses:
        1. Solitary bone cyst.

    2. Fibrous dysplasia.
    3. Desmoplastic fibroma.
 D. Metastatic potential: None.
 E. Radiographic work-up: Plain film appearance
    often leads to biopsy without further work-up
    since it is an unaggressive-looking lesion.
 F. Treatment: Curettage if monostotic.

Cystic Angiomatosis

   Rare, benign, multicentric hemangiomatosis or
lymphangiomatosis, often with severe visceral
involvement.

> **Key Concepts:** Benign lesion that is extremely
> aggressive locally; phleboliths help make the
> diagnosis.

 A. Determinants.
    1. Age: First through third decades.
    2. Soft tissue involvement: Mass with calcified
       phleboliths may be present.
    3. Pattern: Moth-eaten.
    4. Size: Usually several centimeters.
    5. Location: In any bones, most commonly
       femur, pelvis, ribs, humerus, skull, and
       vertebrae.
    6. Zone of transition: Generally narrow, but
       may be indistinct.
    7. Margin: Sclerotic rim, but may be incomplete.
    8. Tumor matrix: Lytic (rarely dense).
    9. Host response: Usually none except cortical
       expansion.
   10. Polyostotic.
 B. Aggressive-looking large, very expanded lesions.
 C. Major differential diagnoses in the usual case of
    polyostotic lesions.
    1. Metastases.
    2. Histiocytosis.
    3. Fibrous dysplasia.
    4. Enchondromatosis (Maffucci's).

D. Metastatic potential: Does not metastasize but is locally aggressive: 50% 5-year survival without visceral involvement; visceral involvement associated with chylous effusions and death.

E. Radiographic work-up: Plain film; otherwise symptomatic.

F. Treatment: Curettage or radiation, depending on extent and site of lesion; recurrence is common, and therapy often is not helpful, especially with visceral involvement.

Massive Osteolysis (Gorham's Disease)

Angiomatosis and regional dissolution of bone with much more extensive destruction than in routine angiomatosis.

A. Most commonly seen in children and young adults.

B. There is usually a *history of trauma*.

C. *Rapid destruction of bone* (not permeative—bone simply disappears) *which spreads contiguously across joints*.

D. No host reaction.

E. Shoulder and hip are most common sites.

F. Radiation may possibly help, but lesions may either stabilize or progress relentlessly.

Glomus Tumor

A rare benign vascular tumor found in the *terminal phalanx* that is *lytic* and well circumscribed. It is usually found in adults, is painful, and is treated by marginal excision or curettage. Its location in the terminal phalanx helps differentiate it from enchondroma and sarcoid lesion of the hand.

B. Vascular Lesions Intermediate or Indeterminate for Malignancy

Hemangiopericytoma

Vascular lesion seen in soft tissues which may erode bone cortex but is extremely rare as an intraosseous lesion. The soft tissue lesion is nonspecific radiographically. The osseous lesion is described here.

A. Determinants:
   1. Age: Fourth or fifth decade.
   2. Soft tissue involvement: Mass common.
   3. Pattern: Permeative or moth-eaten.
   4. Size: Usually less than 5 cm.
   5. Location: Axial skeleton and proximal tubular bones (metaphyseal).
   6. Zone of transition: Wide.
   7. Margin: No sclerotic margin.
   8. Tumor matrix: Lytic.
   9. Host response: Usually none, rare periosteal reaction.
   10. Monostotic.
   11. May be benign, locally aggressive, or malignant.
B. Generally aggressive appearance.
C. Major differential diagnoses: Aggressive lesions.
   1. Fibrosarcoma/MFH.
   2. Intramedullary chondrosarcoma.
   3. Lymphoma.
D. Metastatic potential:
   1. Occasional metastases to lung or bone.
   2. Five-year survival 90%.
E. Radiographic work-up:
   1. Plain film suggests an aggressive lesion and leads to work-up including bone scan, CT or MRI, and chest films. The lesion is nonspecific and extremely rare, so it is unlikely that the diagnosis will be made radiographically.
F. Treatment: Wide resection.

Hemangioendothelioma

Benign or low-grade malignant vascular lesion, often difficult to differentiate histologically from angiosarcoma.

> **Key Concepts:** Moderately aggressive lesion, very often polyostotic and involving the hands or feet.

A. Determinants:
   1. Age: Second or third decade.
   2. Soft tissue involvement: May have a soft tissue mass.
   3. Patterns: Moth-eaten to geographic.
   4. Size: 2 to 5 cm.
   5. *Location: The most important radiographic feature of these tumors. When multicentric, they tend to involve multiple bones of a single extremity, often the hands or feet;* may be cortically based and either diaphyseal or metaphyseal.
   6. Zone of transition: Narrow to intermediate.
   7. Margin: Little or no sclerosis.
   8. Tumor matrix: Lytic.
   9. Host response: Usually none.
   10. *Polyostotic* more common than monostotic.
   11. Other features: Interestingly, the epithelioid hemangioendothelioma has the same histology as intravascular bronchoalveolar tumor (IVBAT).
B. Unaggressive to mildly aggressive appearance.
C. Major differential diagnoses:
   1. Metastases.
   2. Multiple enchondromas.
D. Metastatic potential: Occasionally metastasizes to lungs; interestingly, the *multicentric lesions have a better prognosis* than monostotic ones.
E. Radiographic work-up: Usually a plain film diagnosis.
F. Treatment: Resection if monostotic.

C. Malignant Vascular Tumors

Angiosarcoma

A rare malignant vascular lesion that may be difficult to differentiate histologically from the more benign hemangioendothelioma.

Key Concepts: Aggressive vascular lesion; not uncommonly multifocal; in a young adult, this multifocal character suggests the diagnosis.

A. Determinants:
    1. Age: Fourth and fifth decades; younger if multifocal.
    2. Soft tissue involvement: Mass is common.
    3. Pattern: Permeative.
    4. Size: Usually greater than 5 cm.
    5. Location: Metaphyseal—femur, tibia, humerus, and pelvis are most common sites. *(If the lesion is multifocal, the angiosarcoma may be regional in distribution.)*
    6. Zone of transition: Moderate to wide.
    7. Margin: Usually no sclerotic margin.
    8. Tumor matrix: Lytic.
    9. Host response: Cortex is thinned, expanded, and disrupted without significant host reaction.
   10. *Thirty-eight percent polyostotic.*
B. *Aggressive appearance.*
C. Major differential diagnoses:
    1. If monostotic:
        a. Fibrosarcoma/MFH.
        b. Intramedullary chondrosarcoma.
        c. Lymphoma.
    2. If polyostotic: Metastases.
D. Metastatic potential:
    1. *Highly malignant,* with metastases to lungs or skeleton.
    2. Five-year survival 30% to 50%.
    3. *Prognosis better if the lesions are multifocal.*
E. Radiographic work-up:
    1. Plain film.

2. Bone scan for metastases or multifocal lesion.
3. CT/MRI as needed for diagnosis or treatment plan.
4. Chest x-ray and CT.

F. Treatment: Radical or wide marginal resection.

VI. OTHER CONNECTIVE TISSUE TUMORS

A. Other Connective Tissue Tumors: Benign

Fibromatoses

A heterogeneous group of lesions that have been described with a variety of terms and classifications; the histology of all these lesions is similar. In this chapter, they are divided into *soft tissue* and *intraosseous* fibromatoses.

Soft Tissue Fibrous Tumors.—Tend to be grouped according to time and location of occurrence; they all tend to be *locally infiltrative*, a characteristic mirrored by CT or MRI scans on which *no pseudocapsule is seen* and the tumor often infiltrates through compartmental barriers.

A. *Juvenile aponeurotic fibroma.*
    1. Slowly infiltrative lesion arising in the aponeurotic tissue of the *hands* (palms), *wrist,* and *feet* (soles).
    2. Forms a painless soft tissue mass, usually less than 4 cm in length.
    3. *May calcify*, especially in the *interosseous membrane* of the distal forearm.
    4. Seen in children and adolescents.
    5. *Recurrence* after resection is *common.*
B. *Infantile dermal fibromatosis.*
    1. Infiltrates *extensor surfaces of digits,* presenting as multiple firm nodules attached to skin, tendon, fascia, and periosteum.
    2. Bony erosion rarely occurs.
    3. Appears at 1 to 2 years of age.
    4. *Recurrence* after excision is *frequent.*
C. *Congenital generalized fibromatosis.*

1. Develops in utero.
2. Disseminated fibromatosis involves much of the *musculature and viscera.*
3. *Fatal* within a few months.
4. Another form (congenital multiple fibromatosis) involves only muscle and has a better prognosis. Small well-defined bony erosions are seen occasionally.

D. *Desmoid tumor* (also commonly called *aggressive fibromatosis,* desmoid fibromatosis, or fibrosarcoma grade 1 desmoid type).
   1. Painless *infiltrative soft tissue masses* originating in abdominal or extra-abdominal muscle.
   2. Aggressive local infiltration of adjacent muscles, vessels, nerves, and tendons.
   3. *Bone involvement* is rare (6%) and is extrinsic but *may be spectacular:* huge *frondlike excrescences* may form from a stimulated periosteum, with spicules of bone radiating into a soft tissue mass. Less remarkable *pressure erosion of the cortex* also may be seen.
   4. Often presents in children and may be indolent for long periods.
   5. May be difficult to differentiate histologically from a low-grade fibrosarcoma but rarely, if ever, metastasizes.
   6. The post resection *recurrence rate* is 65% to 75%.

Desmoplastic Fibroma.—The rare form of fibromatosis that is *intraosseous;* same histology as soft tissue fibromatoses.

A. Determinants:
   1. Age: 50% present in *second decade.*
   2. Soft tissue involvement: Generally none.
   3. Pattern: Geographic to moth-eaten, with *cortical expansion* and endosteal erosion.
   4. Size: Variable, but usually greater than 5 cm.
   5. Location: *Central, metaphyseal;* long bones most frequent (around knee); also pelvis and mandible.

　　6. Zone of transition: Usually narrow, but may
　　　be wider, simulating a very aggressive lesion.
　　7. Margin: Sclerotic margin not common.
　　8. Tumor matrix: None (lysis).
　　9. Host response: Generally none.
　10. Monostotic.
B. *Mildly to moderately aggressive, sometimes difficult to*
　differentiate radiographically or pathologically from a
　well-differentiated fibrosarcoma.
C. Major differential diagnoses:
　　1. Fibrosarcoma (if aggressive).
　　2. ABC.
　　3. GCT.
D. Metastatic potential: Not malignant but
　recurrences are very common, though slow (1 to 2
　years).
　　2. Ten to 50% behave as low-grade aggressive
　　　but nonmetastasizing neoplasms, but it is
　　　impossible to predict which on the basis of
　　　radiographic appearance.
E. Radiographic work-up:
　　1. Plain film.
　　2. CT or MRI.
F. Treatment:
　　1. Extensive cryosurgery and curettage.
　　2. With recurrence, wide excision.

Cortical desmoid (periosteal desmoid).—The following
are characteristic of cortical desmoids:

A. *Not a true desmoid*, but fibroblastic proliferation,
　probably secondary to trauma at the insertion of the
　adductor magnus muscle.
B. Causes *erosion of the cortex*, a small soft tissue
　mass, and *exuberent periostitis* that may simulate
　tumor bone formation.
C. *May be misdiagnosed* both radiographically and
　histologically as an *osteosarcoma*.
D. Occurs in the right age group for osteosarcoma
　(15 to 20 years).
E. *Location* is the hallmark of this lesion and leads

to the correct diagnosis: it is always found on the *posteriomedial cortex of the distal end of the femur*, adjacent to the medial femoral condyle.

Lipoma

Tumor arising from fatty tissue, commonly found in soft tissues and rarely intraosseous.

Soft Tissue Lipoma.—Common lesion.

A. Eighty percent are in subcutaneous tissue; others are inter- or intramuscular.
B. Ninety-five percent are solitary.
C. Present as an asymptomatic soft, compressible, and movable mass.
D. On plain film or CT, they are *radiolucent* (fat tissue density) and *well-defined.*
E. Rarely lipomas contain metaplastic cartilage and bone calcification.
F. MR shows a sharply bordered lesion with *high signal intensity on both T1 and T2 weighting.*

Lipomatosis.—The following are characteristic of lipomatosis:

A. Congenital abnormality with multiple lipomas distributed either randomly or symmetrically over body.
B. *Macrodystrophia lipomatosa* is a more localized form, with overgrowth of soft tissues and bone, usually of a hand or foot. This entity belongs in the differential for localized giantism (neurofibromatosis and soft tissue vascular lesions are the other common causes). It is very difficult to handle surgically.

Lipoblastomatosis.—An embryonal fatty tumor seen in young children.

A. It is benign but simulates liposarcoma histologically.
B. Recurrence after surgery is common.
C. Age at onset is the major differentiating factor from liposarcoma, which occurs in adults.

Osseous Lipoma.—Rare fatty lesion of bone.

A. Generally asymptomatic.
B. *Lytic lesion* with a distinct, fine *sclerotic margin* and no destructive change or periosteal reaction.
C. Usually of *fat density on CT* (not completely reliable).
D. May have a *central calcified nidus.*
E. Found in the metaphyses of long bones, especially the *proximal femur*, fibula, and *calcaneus*, (where the osseous lipoma occurs in the triangular region between the major trabecular arcs (same site as SBC).
F. Occasionally the epicenter is *parosteal;* in which case radiographic studies demonstrate the *soft tissue lucent mass adjacent to cortex* plus occasional periosteal reaction; that may take the form of *hyperostosis or may produce large bony spicules radiating* from the periosteum into the lesion.

Giant Cell Tumor of the Tendon Sheath (Pigmented Villonodular Synovitis—PVNS).—The following are characteristic of this lesion:

A. Painless, slow-growing lesion in the tendon sheath, usually of the finger.
B. Patients are usually 30 to 50 years old.
C. Uncertain whether etiology is neoplastic or reactive.
D. Radiographically, a localized soft tissue mass is seen, not centered around a joint, with bony erosion (pressure type) in 10%.
E. Postsurgical recurrence 30%.
F. Intra-articular PVNS, histologically the same lesion but with different radiographic and clinical manifestations, is discussed in Chapter 2.

B. Other Connective Tissue Tumors: Malignant

Malignant Fibrous Histiocytoma (MFH)/ Fibrosarcoma

> **Key Concepts:** Relatively common bone or soft tissue sarcoma in adults; the bony lesion appears permeative and aggressive; the soft tissue lesion may appear misleadingly "encapsulated" and benign.

MFH is a relatively recently described lesion that encompasses many tumors previously felt to be distinct entities; many lesions previously diagnosed as fibrosarcomas would now be designated MFH.

Because fibrosarcoma and MFH usually can be distinguished histologically, but not radiographically, the two are described together in this chapter.

Either MFH or fibrosarcoma *may originate in the skeleton*, where it is usually an aggressive sarcoma, *or in the soft tissues* (the most common soft tissue sarcoma in adults), where it may be either low-grade or highly aggressive.

The *soft tissue MFH/fibrosarcomas*, like many soft tissue sarcomas, may have a reactive pseudocapsule that makes it *appear encapsulated on CT, MRI, and at surgery*, when the lesion may "shell out" easily. It is important not to be misled by this pseudocapsule into treating it as a nonaggressive lesion: tumor cells invariably are found outside the margin of the reactive pseudocapsule.

Osseous fibrosarcomas and MFH *may be primary bone tumors or may arise secondarily*. Either may be found secondary to previous *irradiation, Paget's disease degeneration, or dedifferentiation of a chondrosarcoma*. MFH may, in addition, *arise in a* previous *bone infarct*, possibly secondary to chronic repair processes.

Osseous MFH and fibrosarcoma are described below:

A. Determinants:
   1. Age: Second through seventh decades, but fifth or sixth decades is most common.
   2. Soft tissue involvement: Large soft tissue mass is common.
   3. Pattern: *Permeative* or moth-eaten.

4. Size: Greater than 5 cm.
5. Location: *Central* (but may be eccentric); diametaphyseal; 75% in *long tubular bones*, especially around the knee, humerus, and pelvis.
6. Zone of transition: Wide.
7. Margin: No sclerotic margin.
8. Tumor matrix: Lytic lesion, though dystrophic calcification may be seen in 15%; serpiginous calcification may also be seen in residual portions of bone infarcts that have degenerated to MFH.
9. Host response: Periosteal reaction is variable but often present.
10. Usually monostotic; multicentric lesions are extremely rare and more likely represent bone metastases.
B. *Highly aggressive* lesions.
C. Major differential diagnoses:
   1. Lymphoma.
   2. Intramedullary chondrosarcoma (especially if calcification is present).
D. Metastatic potential:
   1. Very few are low-grade lesions (5-year survival 85% to 95%).
   2. Most are high-grade, with a 5-year survival of 25% and metastases involving lung, bone, lymph nodes, and viscera.
E. Radiographic work-up:
   1. Plain film: Suggests diagnosis and limited differential of aggressive lesion.
   2. Bone scan for metastases.
   3. Chest film and CT for metastases.
   4. CT or MRI for determination of tumor extent and surgical planning.
F. Treatment:
   1. Low-grade: Wide excision; value of chemotherapy is debated.
   2. High-grade: Wide to radical excision and chemotherapy.

Liposarcoma

A. Liposarcoma of bone is extremely rare. It is aggressive in appearance as well as clinically and is nonspecific radiographically.

B. Soft tissue liposarcoma is the *second most common soft tissue sarcoma.*

C. It may be well-differentiated or a high-grade lesion.

D. Third, fourth, or fifth decade.

E. Insidious growth, often asymptomatic, so may be large at diagnosis.

F. Most commonly located in the *buttocks, thigh,* lower leg, and retroperitoneum.

G. On plain film or CT, *may have fat density if the lesion is well-differentiated. A higher grade lesion is more cellular and more often of nonspecific soft tissue density.*

H. Occasionally may contain *dystrophic calcification,* bone, or cartilage.

I. MRI shows a high signal intensity T2 (as in nearly all soft tissue lesions) and a variable signal intensity T1 (most other soft tissue lesions except lipoma have a low signal intensity T1).

J. The lesion often elicits formation of a *reactive pseudocapsule;* this is *easily misinterpreted on CT, MRI, and at surgery as true encapsulation* and the lesion is "shelled out," invariably leaving *residual tumor at the margins* since the lesion is never truly encapsulated; thus recurrence with this type of excision is common and prognosis is poor, with metastatic disease to lungs and viscera.

K. Adequate therapy is wide excision, often with chemotherapy, often combined with radiation pre- or postoperatively.

Synovial Cell Sarcoma

A. Soft tissue sarcoma of synovial origin.

B. Presents with painful swelling but relatively slow growth.

C. Age: 15 to 35 years.

D. Associated with tendon, tendon sheath, and bursa and therefore may be *remote from joint;* fewer than 10% occur in the joint capsule.

E. Most common in the *lower extremity*, especially about the knee.

F. *Calcification is seen in 20% to 30%.*

G. Adjacent bone erosion or periosteal reaction is rare (15%).

H. Radiographic appearance is, therefore, nonspecific; as in other soft tissue sarcomas, a misleading *pseudocapsule* may be seen, but islands of tumor are found peripheral to it.

I. Recurrence is common unless the lesion is treated initially with wide excision.

J. Metastasizes to lung.

Other Connective Tissue Sarcomas

Radiographically, these are all nonspecific; they tend to be masses of soft tissue density, sometimes containing calcification and often with an apparent pseudocapsule.

Malignant Mesenchymoma.—Contains elements of osteosarcoma and liposarcoma.

Rhabdomyosarcoma.—Common soft tissue tumor of muscle origin; poor prognosis, with metastases to lung and lymph nodes. Embryonic type occurs in children, primarily in the head and neck and genitourinary tract. The alveolar, or pleomorphic, types occur in adults, usually in the extremities.

Malignant Schwannoma.—In the extremities, associated with neurofibromatoses.

Extraskeletal Osteosarcoma (see Osteosarcoma, above chap).—Usually has osteoid matrix.

Extraskeletal Chondrosarcoma.—Chondroid matrix.

VII. OTHER TUMORS
Chordoma

A low-grade malignant neoplasm that arises from *notocord remnants.*

> **Key Concepts:** *Locally aggressive;* because of its origin, it is found only in the *sacrum, clivus, and spine;* recurrence is common.

A. Determinants:
  1. Age: In sacrum, sixth and seventh decades; in clivus or spine, fourth and fifth decades.
  2. Soft tissue involvement: *Local soft tissue mass.* In the sacrum the soft tissue mass enlarges anteriorly into the pelvis and may be very large by time of discovery; in the spine there may be a posterior mass with epidural compression.
  3. Pattern of bone destruction: Usually geographic.
  4. Size of lesion: Usually larger than 5 cm in sacrum but smaller in spine.
  5. Location:
     a. *Fifty percent in sacrum (conversely, 40% of sacral tumors are chordomas).*
     b. Thirty-five percent in clivus.
     c. Fifteen percent in spine: Body rather than posterior elements; usually lumbar.
  6. Zone of transition: Narrow.
  7. Margin of lesion: Usually sclerotic.
  8. Tumor matrix: No matrix produced, but calcific debris may be present.
  9. Host response: No periosteal reaction.
  10. Monostotic.
B. Aggressiveness of lesion: *Although extensive local bone destruction and soft tissue mass are seen, the time course is often so slow that the lesion acquires a sclerotic rim and has a narrow zone of transition, making it appear less aggressive.*
C. Major differential diagnoses:
  1. In sacrum:
     a. GCT.
     b. Chondrosarcoma.
     c. Plasmacytoma.

  2. In vertebral body:
     a. Metastatic disease.
     b. Myeloma.
  3. In clivus:
     a. Chondrosarcoma.
     b. Metastatic.
D. Metastatic potential.
  1. *Twenty-five percent have distant metastases (to lung), but these are often very late.*
  2. *More commonly, patients have significant morbidity and mortality from local recurrence and associated complications.*
  3. Five-year survival 50%.
  4. Ten-year survival 28%.
E. Radiographic work-up:
  1. Plain film suggests diagnosis.
  2. CT or MRI ideal to diagnose local extent.
F. Treatment:
  1. Early wide resection, if possible.
  2. *Local recurrence 80% if only marginal resection is accomplished.*
  3. Radiation seems to be palliative for recurrence but does not affect survival.
  4. Chemotherapy is not helpful.

Adamantinoma (angioblastoma)

Rare lesion of unknown pathogenesis (contains elements of squamous, alveolar, and vascular tissue) and generally low-grade malignancy.

> **Key Concepts:** Location in the tibial diaphysis may be its most distinctive feature; appearance ranges from unaggressive to moderately aggressive; may have satellite foci and may be malignant.

A. Determinants:
  1. Age: Most commonly fourth or fifth decade but may occur in adolescence.

2. Soft tissue involvement: Rare early, but a soft tissue mass may develop as the lesion becomes more aggressive locally.
3. Pattern of bone destruction: Geographic.
4. Size of lesion: Depending on the stage, may be greater than 5 cm.
5. Location:
   a. *This is the most characteristic aspect of adamantinoma: 90% are found in the tibia.* Other long bones are affected rarely.
   b. It is usually *diaphyseal (middle one third)* and *eccentric* (often cortically based initially).
6. Zone of transition: Geographic (though the more aggressive lesions may be moth-eaten).
7. Margin of lesion: Usually has reactive sclerosis.
8. Tumor matrix: None.
9. Host response: No periosteal reaction unless lesion is very aggressive.
10. Monostotic, but *may have satellite foci adjacent to the parent lesion or even in the adjacent fibula.*
11. This lesion is distinct from adamantinoma (ameloblastoma) of the jaw.

B. Aggressiveness of lesion: *Appearance ranges from unaggressive to moderately aggressive.*

C. Major differential diagnoses:
   1. Fibrous dysplasia.
   2. NOF.
   3. Vascular lesion (especially when satellite lesions are present).

D. Metastatic potential:
   1. Malignant (20% metastasize to lung, lymph nodes, or skeleton) but generally low-grade, so the lesion may be present for several years prior to metastasis.
   2. Five-year survival 60%.
   3. Ten-year survival 40%.

E. Radiographic work-up:
   1. Plain film: *Adamantinoma should be suggested by a moderately aggressive midtibial lesion.*

2. CT or MRI: To evaluate local extent and fibular involvement. This is important since the lesion is usually low-grade, and the poor survival statistics are likely to relate to underestimated local extension and inadequate initial treatment.
3. Chest film.

F. Treatment:
1. Ideally, wide excision should be performed initially.
2. Frequently the initial treatment is an inadequate curettage or cryosurgery.

Neurofibroma

See Chapter 5, section III.

VIII. TUMORLIKE LESIONS
Solitary Bone Cyst (SBC)

Also called simple or unicameral bone cyst; common fluid-filled bone lesion of childhood. It is asymptomatic unless fractured (Fig 1–11).

> **Key Concepts:** Central geographic lesion, common in young children; metaphyseal but may migrate to diaphysis as it matures; proximal humerus is most common site; very high recurrence rate after simple curettage.

A. Determinants:
1. Age: *First or second decade;* may uncommonly be seen in adults.
2. Soft tissue involvement: None.
3. Pattern of bone destruction: Geographic oval lesion with its long axis parallel to the long axis of the bone; may appear multilocular.
4. Size of lesion: May be 5 cm or larger.
5. Location: *Proximal humerus (50%) and proximal femur (20%) most common; SBC is metaphyseal,*

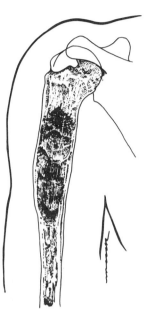

FIG 1–11.
Solitary bone cyst: mildly expansile central lesion that is lytic and located in the metaphysis when it is active. The cortex is intact and there is no soft-tissue mass. Because it is asymptomatic unless fractured, it is often detected serendipitously on a film taken for other reasons.

abutting the epiphyseal plate, but migrates into the diaphysis (normal bone grows away from it). An SBC that has migrated from the growth plate may be less active and, thus, less prone to recurrence than those that remain at the growth plate; however, activity seems to correlate better with patient age (recurrence is twice as likely in a patient under age 10 than in an older patient). An SBC in an older patient is rare and is found in unusual locations, such as the iliac wing or calcaneus.

6. Zone of transition: Narrow.
7. Margin of lesion: *SBCs are mildly expansile and thin the cortex, but have a fine sclerotic rim.*
8. Tumor matrix: None, but rarely one may see a *fallen fragment sign:* after pathologic fracture of the cyst, a fracture fragment may be displaced inferiorly in the fluid-filled cyst.
9. Host response: No periosteal reaction unless there is a pathologic fracture.
10. Monostotic.

B. Unaggressive in appearance.
C. Major differential diagnoses:
1. Fibrous dysplasia.
2. Eosinophilic granuloma.
3. Aneurysmal bone cyst (more eccentrically located).

D. Metastatic potential: None.
E. Radiographic work-up: Plain film diagnosis.
F. Treatment:
1. *Curettage and bone graft has a recurrence rate of 35% to 50%.*
2. *Steroid injection* (after proving diagnosis by aspiration of fluid and pressure measurements) or cryosurgery show lower recurrence rates.
3. Occasionally may heal spontaneously following multiple fractures.
4. Surgery performed on a lesion adjacent to the epiphysis it may cause acceleration or arrest of growth.

Aneurysmal Bone Cyst (ABC)

An expansile, highly vascular lesion with blood-filled cystic cavities. It is associated with a preexisting osseous lesion (chondroblastoma, fibrous dysplasia, GCT, osteoblastoma, NOF) *in 30% to 50% of cases. Trauma* is also a common feature; one theory of the pathogenesis of ABC is that it is a vascular anomaly induced by either trauma or the precursor lesion. It is hypothesized that rapid expansion of the ABC obliterates the precursor lesion in some cases (Fig 1–12).

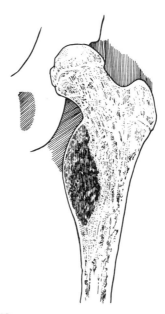

FIG 1–12.
Aneurysmal bone cyst: very expansile, eccentrically located metaphyseal lytic lesion with an extremely thin cortical rim.

> *Key Concepts:* Very expansile, eccentric, lytic metaphyseal lesion with narrow zone of transition; usually under 25 years of age; may be very rapidly progressive and simulate neoplasm.

A. Determinants:
   1. Age: First through third decades; *70% between 5 and 20 years.*
   2. Soft tissue involvement: Soft tissues are displaced by the rapidly expanding lesion.
   3. Pattern of bone destruction: Geographic.
   4. Size of lesion: Often greater than 5 cm.
   5. Location: *Metaphyses of long bones, spine (posterior elements), pelvis;* it is *eccentrically* located.
   6. Zone of transition: Narrow.
   7. Margin of lesion: Fine sclerotic rim; the margin outside the bone may not be seen by plain film but more commonly can be seen with CT.
   8. Tumor matrix: None.
   9. Host response: Aggressive recurrence may elicit periosteal reaction, but usually there is no host response.
   10. Monostotic.
   11. Other features: *May be very rapidly progressive* and elicit periosteal reaction, which may be mistaken for a malignant tumor.
B. Usually the appearance is unaggressive; rapidly progressive or recurrent lesions may appear more aggressive.
C. Major differential diagnoses:
   1. NOF.
   2. Fibrous dysplasia.
   3. Osteoblastoma (in posterior elements of spine).
D. Metastatic potential: None, but the underlying lesion may occasionally be a sarcoma; therefore, extensive sampling of solid areas should be

performed to determine the precursor lesion.
E. Radiographic work-up: Plain film diagnosis usually. If more aggressive, CT or MRI will help further define the lesion; a fluid level may be present.
F. Treatment: Curettage and possibly cryosurgery (there is a *50% recurrence rate* following curettage). Low-dose radiation may be used only for surgically inaccessible lesions.

Nonossifying Fibroma (NOF)/Benign Fibrous Cortical Defect (BFCD)

Histologically identical, cortically based lesions that are not neoplasms but may be secondary to epiphyseal plate defects that migrate away from the plate with growth. It is estimated that *BFCD occurs in 30% to 40% of children* over 2 years of age; they are seen in adults infrequently so must heal *spontaneously*. Occasionally BFCD may enlarge, forming an NOF. The lesions are asymptomatic and are so radiographically specific that they are among the "leave me alone" lesions that should be ignored unless symptomatic.

> *Key Concepts:* Cortically based metaphyseal ly-tic lesions with sclerotic border; usually lower extremity long bones; both lesions may sclerose in during healing phase.

A. Determinants:
1. Age: *First or second decade* (95% under 20 years).
2. Soft tissue involvement: None.
3. Pattern: *Geographic oval lesion,* parallel to the long axis of the bone. *NOF is expansile and may have pseudotrabeculations.*
4. Size:
a. BFCD less than 2 cm.
b. NOF greater than 2 cm, occasionally huge.
5. Location: *80% in the long bones of the lower*

extremity; metaphyseal and cortically based (an oblique film may be necessary to demonstrate this). *NOF starts in cortex but may enlarge to involve the intramedullary region and even appear central in thin bones (fibula, ulna).*

6. Zone of transition: Narrow.
7. Margin: Sclerotic border.
8. Tumor matrix: None, but *when healing spontaneously one may see dense sclerotic bone formation.*
9. Host response: No periosteal reaction.
10. *Usually monostotic (75%),* but the lesion is so common that polyostotic lesions are seen not infrequently.
11. Other features: NOF may be associated with neurofibromatosis.

B. *Unaggressive appearance.*
C. Major differential diagnoses:
   1. BFCD should be recognized without having to formulate a differential diagnosis.
   2. NOF differential includes chondromyxoid fibroma and ABC, and occasionally Brown tumor of HPTH.
D. Metastatic potential: None.
E. Radiographic work-up: Plain film should be sufficient.
F. Treatment:
   1. *BFCD: "Leave me alone"*; natural history is to heal in.
   2. NOF: If treatment is indicated secondary to pain or pathologic fracture, curettage with bone chip packing is performed. Otherwise, NOF gradually migrates away from the metaphysis and is remodeled or ossifies in.

Histiocytosis X

Spectrum of diseases, all with histiocytic infiltration of tissues and *aggressive bone lesions. Eosinophilic granuloma (EG), the most common (60% to 80%) and mildest form, is described below.*

> **Key Concepts:** Consider this lesion when considering round cell tumors. EG may appear as permeative and aggressive as Ewing's (though a soft tissue mass is usually not as large); may develop and enlarge extremely rapidly; often polyostotic; skull, spine, pelvis, and femur are the most common sites.

A. Determinants:
  1. Age: First through third decade, but peaks from *5 to 10 years.*
  2. Soft tissue involvement: Soft tissue mass is common, especially in skull lesions.
  3. Pattern: *Moth-eaten or permeative;* as lesion resolves, becomes more geographic.
  4. Size of lesion: 1 to 5 cm.
  5. Location:
     a. *Skull (calvarium) 50%.*
     b. *Axial skeleton 30%;* pelvis; also involves vertebral body and occasionally an adjacent vertebral body in compression fracture *(vertebra plana)* pattern. Posterior elements and discs are intact.
     c. *Long bones 20%;* femur most common. The lesions are usually central and metadiaphyseal.
  6. *Zone of transition: Wide,* especially *in appendicular lesions; skull lesions* have a *narrower* zone of transition, that of a *punched-out* appearance. In addition, the skull lesions often have nonuniform involvement of the inner and outer skull tables, giving a *beveled edge* appearance.
  7. Margin of lesion: No sclerotic margin unless in reparative phase.
  8. Tumor matrix: No matrix, but occasionally a fragment of bone is left centrally in the lesion, resembling a *"sequestrum."* Sclerosis *may be seen during healing phases.*

9. Host response: *Periosteal reaction is common.*
10. Monostotic most commonly; 10% develop polyostotic disease within 6 months of developing the first lesion.
11. Other features: EG involves only a *single organ system;* it is a *painful lesion.* Rarely, it presents with fever and elevated sedimentation rate, simulating infection. It is very aggressive: the time course of bone destruction may be even shorter than that is seen with infection or tumor.

B. *Highly aggressive lesion, both in time course and appearance.*
C. Major differential diagnoses:
   1. Ewing's sarcoma.
   2. Lymphoma.
   3. Infection.
   4. Metastatic disease.
D. Metastatic potential: None; may have multifocal EG without extraskeletal involvement.
E. Radiographic work-up:
   1. Plain film: Suggests diagnosis and differential.
   2. Bone scan: To determine whether it is polyostotic; other lesions may be in the skull or vertebral bodies and lead to the diagnosis of EG.
F. Treatment:
   1. Many therapeutic regimens have been used. The healing rate seems to be similar for all treatments:
      a. No therapy.
      b. Curettage.
      c. Wide excision.
      d. Low-dose radiation.
      e. Intralesional steroid injection.
   2. *Each of these methods seems to lead to the same rate of recurrence or reconstitution* (including regaining partial height in vertebra plana lesions).
   3. Therefore, therapy is often reserved for

specific clinical indications, including the painful lesion.

Hand-Christian-Schuller Disease

1. *Chronic disseminated form* of histiocytosis; 15% to 40% of histiocytoses.
2. Involves skeletal, reticuloendothelial, and other visceral sites.
3. The skeletal lesions have the same appearance as in EG.
4. Manifest by age 5 to 10.
5. Variable prognosis, with high morbidity and 10% mortality.

Letterer-Siwe Disease

1. *Acute fulminant* form of histiocytosis.
2. Ten percent of cases.
3. Involves the skin, liver, spleen, lymph nodes, and skeleton.
4. Skeletal lesions may not be focal but are diffuse and poorly defined (as in myeloma or leukemia).
5. Manifest before age 2.
6. Most cases are fatal, though a few may convert to Hand-Christian-Schuller disease and the patients survive.

Fibrous Dysplasia

A hamartomatous *fibro-osseous metaplasia* (a fibrous stroma with islands of osteoid and woven bone) that is relatively *common*.

> **Key Concepts:** The radiographic manifestations all include expansile lesions but differ, depending on the region of body involved. Rib and long bone lesions are mildly expansile, pelvic lesions are bubbly and may be very large, and base of skull lesions are expansile but densely sclerotic. In the femur, a shepherd's crook deformity is common; when polyostotic, tends to be ipsilateral.

A. Determinants:
  1. Age: 10 to 70, but most often recognized in second or third decade; the polyostotic form is usually recognized before age 10 years.
  2. Soft tissue involvement: None.
  3. Pattern of bone destruction: Geographic, expansion with cortical thinning at all sites but may become extremely expansile and bubbly in the pelvis.
  4. Size: Usually greater than 5 cm.
  5. Location: Found in any bone, but vertebral localization is uncommon. Most *common areas of involvement include the tubular bones (central, metadiaphyseal lesions of femur or tibia), ribs (the most common benign lesion of ribs), pelvis, and skull (frontal, sphenoid, maxillary, and* ethmoid bones).
  6. Zone of transition: Narrow.
  7. Margin: Fine *sclerotic rim*.
  8. Tumor matrix: *Lesions range from completely lucent to a more opaque (ground glass) density, depending on the amount of osteoid or woven bone present). Base of skull lesions often are densely sclerotic.*
  9. Host response: No periosteal reaction in the absence of fracture.
  10. Seventy percent are monostotic.
  11. Other features:
     a. *Polyostotic lesions tend to be more aggressive; 90% are unilateral in distribution.*
     b. Three percent of patients with fibrous dysplasia have *Albright's syndrome*—polyostotic bone lesions, cafe-au-lait spots, and precocious puberty.
     c. Complications of fibrous dysplasia include *fracture (in 40%)* and physical deformity *(shepherd's crook varus deformity of the* proximal femur, *leg length discrepancy,* bowing of long bones).
     d. *Cherubism,* the typical facial deformity, is

caused by expanding lesions of the paranasal sinuses and mandible.

e. *Osteofibrous dysplasia-pseudoarthrosis* of the tibia: In infants to age 5; is a form of fibrous dysplasia resulting in bowing and fracture through the lesion followed by development of pseudoarthrosis; it usually heals with immobilization.

B. Aggressiveness of lesion: Unaggressive in appearance (although polyostotic lesions may appear more aggressive than monostotic lesions).

C. Major differential diagnoses:
  1. For monostotic tubular bone: SBC.
  2. For polyostotic lesions: Ollier's disease or metastases.
  3. For rib lesion: EG, Ewing's, metastasis.
  4. For base of skull lesion: Hyperostosis may suggest meningioma; Paget's may also be in the differential.

D. Metastatic potential:
  1. *Most lesions remain quiescent throughout life, neither improving or resolving;* only 5% continue to enlarge after skeletal maturity.
  2. Malignant transformation (to fibrosarcoma or osteosarcoma) has been reported rarely.

E. Radiographic work-up:
  1. Plain film diagnosis.
  2. If necessary, bone scan will detect polyostotic lesions.

F. Treatment: Symptomatic only, usually osteotomies to reduce deformities.

Brown Tumor of HPTH

Unaggressive-looking lytic lesion generally occurring in the presence of other manifestations of HPTH, especially subperiosteal resorption. (See Chapter 4, section III.)

Myositis Ossificans

Posttraumatic bone formation which, in the early stages, may be adjacent to bone and elicit periosteal reaction, simulating osteosarcoma. Time course and zoning within the lesion are characteristic and help make the correct diagnosis; for further details, see Myositis in Chapter 3, section XIII.

IX. METASTATIC DISEASE OF BONE

A. Frequency of occurrence:
   1. Osseous metastases eventually occur in 20% to 35% of extraskeletal malignancies.
   2. Metastases to bone are significantly more common (25:1) than primary bone tumors.
   3. Eighty percent of bone metastases are from primary breast, prostate, lung, or kidney carcinomas; other common primaries include GI, thyroid, and round cell (lymphoma, neuroblastoma).
B. Imaging modalities:
   1. Plain films and radionuclide bone scan have complementary roles.
   2. Bone scans are highly sensitive compared to plain films: 10% to 40% of metastatic lesions are abnormal on bone scan but normal radiographically. On the other hand, fewer than 5% of metastatic lesions are normal on bone scan but abnormal by plain radiography.
   3. Bone scan specificity is very poor. Abnormalities on radionuclide scan may be due to tumor, trauma, arthritis, or infection; plain film is more specific and differentiates among these possibilities.
   4. In patients with diffuse breast or prostate metastases bone scans may have a "super scan" pattern (diffuse excessive bone uptake with no kidney uptake).
   5. While bone scan and plain film remain the major imaging modalities, CT and MRI can demonstrate marrow infiltration and destruction; they are, however, nonspecific

and are rarely cost-effective diagnostic
modalities for metastases.
6. Angiography and embolization may decrease
morbidity in highly vascular metastases
(particularly kidney).
C. Plain radiographs are used primarily for:
1. Improving specificity (diagnosis).
2. Assessment of therapeutic success.
3. Evaluation for signs of impending pathologic
fracture that would indicate prophylactic
therapy:
a. Lesions 2.5 cm or larger.
b. Fifty percent cortical width destruction.
c. Pain.
D. Appearance of metastases:
1. Generally have a moth-eaten pattern with an
ill-defined zone of transition, no sclerotic
margin, often no periosteal reaction, and a
small soft tissue mass. Thus, they appear
moderately to highly aggressive.
2. Occasionally a metastasis may present as a
geographic, bubbly, expansile mass. The
primary in these cases is usually kidney or
thyroid.
3. They are generally polyostotic (only 10% of
metastases are solitary, usually from a kidney
or thyroid primary).
4. The density of metastases is variable.
a. Purely lytic metastases (in descending
order of frequency) include: Lung, kidney,
breast, thyroid, GI, neuroblastoma.
b. Mixed lytic and blastic metastases include:
Breast, lung, prostate, bladder,
neuroblastoma.
c. Blastic metastases include: Prostate,
breast, bladder, GI tract (stomach,
carcinoid), lung (oat cell),
medulloblastoma.
d. Changing patterns of density may reflect
healing due to therapy, progression of

destruction, or radiation osteonecrosis.
E. Location:
   1. Usually involve marrow spaces centrally (cortex-based metastases are most commonly lung or breast).
   2. Eighty percent are located in the axial skeleton (ribs, pelvis, vertebrae, skull, proximal humerus, and femur). Acral lesions—distal to the elbows or knees—are usually secondary to primary lung tumor.
   3. Metastases involving the spine are very common (found in 38% of malignancies at autopsy, though fewer are detected radiographically). Vertebral metastases may appear as nonspecific compression fractures but more commonly have the following features: involvement of a pedicle, focal destruction, focal soft tissue mass, intact disc.
   4. Lesser trochanter fractures should be considered pathologic until proven otherwise.
   5. Breast metastases in the femoral head and neck usually also involve the acetabulum (although not radiographically demonstrable). Total hip arthroplasties rather than bipolar endoprostheses should be strongly considered for these patients if prophylactic or salvage surgery is to be done.
F. Differential diagnosis:
   1. In adults:
      a. Multiple myeloma.
      b. Radiation osteonecrosis (may appear highly destructive, tend to involve bones adjacent to one another and is restricted to a radiation port).
      c. Sclerosing dysplasias.
      d. Posttraumatic osteolysis of the pubis may simulate aggressive metastatic disease. The trauma may be remote and forgotten. The histologic appearance may be aggressive owing to fracture healing; thus,

trauma should be considered as an
etiology for a lytic pubic lesion in an
elderly patient.
2. In children:
   a. Leukemia.
   b. Disseminated osteomyelitis.
   c. Histiocytosis.
   d. Nonaccidental trauma (e.g., abuse):
      metaphyseal irregularities and periosteal
      reaction.
   e. Metabolic stress with deossification: affects
      rapidly growing parts of bone.
   f. Fibrous dysplasia or Ollier's disease: tends
      to be unilateral in distribution.

REFERENCES

1. Enneking WF: Staging of musculoskeletal
   neoplasms. *Skeletal Radiology* 1985; 13:183–194.
2. Kirchner P, Simon M: The clinical value of bone
   and gallium scintigraphy for soft tissue sarcomas
   of the extremities. *J Bone Joint Surg* 1984;
   66A:319–327.
3. Pettersson H, Gillespy T, Hamlin D, et al:
   Primary musculoskeletal tumors: Examination
   with MR imaging compared with conventional
   modality. *Radiology* 1987; 164:237.
4. Goorin A, Abelson H, Frei E: Osteosarcoma:
   Fifteen years later. *N Engl J Med* 1985; 313:1637–
   1643.
5. Simon M: Causes of increased survival of
   patients with osteosarcoma: Current
   controversies. *J Bone Joint Surg* 1984; 66A:306–310.
6. Huvos A: Osteogenic sarcoma of bones and soft
   tissues in older persons. *Cancer* 1986; 57:1442–
   1449.
7. Schwartz H, Zimmerman N, Simon M, et al.: The
   malignant potential of enchondromatosis, *J Bone
   Joint Surg* 1987; 69:269–274.

BIBLIOGRAPHY

Mirra J: *Bone Tumors: Diagnosis and Treatment.*
  Philadelphia, J. B. Lippincott, 1980.
Enneking W: *Musculoskeletal Tumor Staging, vols. I and
  II.* New York, Churchill Livingstone, 1983.
Hudson T: *Radiologic-Pathologic Correlation of
  Musculoskeletal Lesions.* Baltimore, Williams &
  Wilkins, 1987.

2

Arthritis

GENERALIZATIONS

The classic appearance of most arthritides in the chronic stages makes them relatively easy to distinguish radiographically. It is in the early stage of disease that accurate diagnosis may be difficult. A monoarticular arthritis could easily result from trauma, infection, a crystal-induced arthropathy, early rheumatoid arthritis (RA), seronegative arthritis, or osteoarthritis (OA).

Several parameters, when used in combination, usually lead to accurate diagnosis of an early arthropathy. These parameters include clinical evaluation, epidemiologic factors such as age and sex of the patient, distribution of the arthropathy (involved joints), general appearance (erosive versus productive bony changes), and, occasionally, laboratory tests. Because these parameters are usually quite reliable for the various arthropathies, the arthritis section of this handbook is organized to highlight them.

A. Definition.
B. Epidemiology: Age, sex.
C. Clinical signs: Pain, stiffness, swelling, altered range of motion (ROM), deformity.
D. Pertinent laboratory tests (often not necessary but may be confirmatory).
E. Extra-articular manifestations.

F. General radiographic description.
   1. Soft tissue alterations.
   2. Abnormal calcifications.
   3. Bone density.
   4. Cartilage destruction.
   5. Erosive versus productive bone changes; In general, the erosive arthropathies have an initial inflammatory stage which produces pannus (inflammatory granulation tissue), which destroys cartilage and bone by means of lytic enzymes. The classic example of an erosive arthropathy is RA. At the other end of the spectrum, OA also involves cartilaginous and subchondral bone destruction, but abnormal mechanical forces combined with host reactive processes lead to productive changes (osteophyte formation, subchondral sclerosis, and buttressing). Many other arthropathies (seronegative, crystal-induced, and erosive osteoarthritis) generally fall between these two ends of the spectrum, often demonstrating both erosive and productive changes.
   6. Subchondral cysts.
   7. Periostitis or enthesopathy ("whiskering" periostitis at the site of attachment of a ligament or tendon).
   8. Ankylosis of joint.
   9. Ligamentous abnormality: Rupture, laxity, or contracture leading to subluxation, dislocation, or instability.
G. Joints most commonly affected: This distribution is a major diagnostic guide, with specific comments regarding individual sites of involvement.
H. Bilateral symmetry.
 I. Other features.
 J. Differential diagnosis.
K. Suggested survey films for diagnosis of early disease.

In this handbook, the arthropathies have been organized according to the American Rheumatism Association (ARA) classification, with modifications and deletions (either entities that are rare, do not have radiographic abnormalities, or are discussed elsewhere in the handbook):

I. Polyarthritis of unknown etiology.
   A. Rheumatoid arthritis (RA).
   B. Juvenile rheumatoid arthritis (JRA).
   C. Ankylosing spondylitis (AS)/inflammatory bowel disease (IBD) spondylitis.
   D. Psoriatic arthritis.
   E. Reiter's syndrome.
II. Connective tissue disorders.
   A. Systemic lupus erythematosus (SLE).
   B. Progressive systemic sclerosis (scleroderma).
   C. Polymyositis/dermatomyositis.
   D. Amyloidosis.
III. Rheumatic fever.
IV. Osteoarthritis (OA).
V. Neuropathic arthropathy.
VI. Arthropathy secondary to biochemical abnormalities.
   A. Gout.
   B. Calcium pyrophosphate dihydrate (CPPD) crystal deposition disease.
   C. Hemochromatosis.
   D. Wilson's disease.
   E. Calcium hydroxyapatite (HA) deposition disease.
   F. Ochronosis.
VII. Miscellaneous disorders.
   A. Pigmented villonodular synovitis (PVNS).
   B. Synovial chondromatosis.
   C. Osteochondroses/osteochondritis dessicans/spontaneous osteochondritis.
   D. Hypertrophic osteoarthropathy.
   E. Avascular necrosis (AVN).

F. Diffuse idiopathic skeletal hyperostosis
(DISH; Forrestier's disease).
G. Transient regional osteoporosis.

For a much more detailed review of the pathologic
basis of these disease processes, the reader is referred
to the impressive compilation of data and bibliographies
in Resnick D, Niwayama G: *Diagnosis of Bone and Joint
Disorders*, ed 2. Philadelphia, W.B. Saunders, 1988.

I. POLYARTHRITIS OF UNKNOWN ETIOLOGY
Rheumatoid Arthritis (RA)

> *Key Concepts:* Erosive arthropathy; synovitis
> and osteoporosis; bilaterally symmetric; carpus,
> metacarpo- and metatarsophalangeal joints
> (MCPs, MTPs), elbows, shoulders (rotator cuff
> tear), knees (valgus), hips (protrusio), upper
> cervical spine pathology (facet erosions,
> atlantoaxial impaction or subluxation).

A. Definition: A common arthritis of unknown
etiology that causes synovial inflammation and
articular destruction that is invariably
polyarticular.
B. Epidemiology:
1. Age: Young or middle-aged.
2. Sex: *Females* more commonly affected (2 or
3:1).
C. Clinical signs:
1. Symptoms may be chronic or episodic.
2. Early morning stiffness.
3. Pain (due to capsular distension).
4. Periarticular muscle wasting, but boggy
synovial swelling.
5. Tendon contractures and rupture result in
several characteristic deformities (see G,
below).
D. Laboratory tests:

1. *Rheumatoid factor* (RF) may be negative early in the disease process but eventually becomes positive in 90% to 95% of cases; elderly patients may be false positive.
2. Erythrocyte sedimentation rate (ESR) is elevated and tends to parallel disease activity.

E. Extra-articular manifestations.
   1. Subcutaneous or tendon sheath nodules.
   2. Tenosynovitis or bursitis.
   3. Erosions at entheses.
   4. Irregularities at discovertebral junction (especially in the cervical spine).
   5. Pleural effusion.
   6. Rheumatoid pulmonary nodules.
   7. Diffuse interstitial pneumonitis.

F. General radiographic description.
   1. Soft tissue alterations: *Fusiform swelling* around joints secondary to effusions and synovitis; in addition, *large synovial cysts* may form that communicate with the joint.
   2. Abnormal calcifications: None.
   3. Bone density: *Osteoporotic*, owing to a combination of hyperemia and disuse; in early disease, the osteoporosis may be only juxtaarticular; later, generalized; thinning of the subchondral cortex may be seen early as a *dot-dash pattern.*
   4. Cartilage destruction: Joint space initially may appear wide due to joint distension. Subsequently the *cartilage is destroyed in a uniform pattern,* leading to joint space narrowing.
   5. *Erosive changes:* Bone destruction initially is *marginal, at the bare areas* (within the joint capsule but not protected by cartilage). *Later,* with cartilage destruction, *subchondral erosions occur.* Productive bone changes are rare, but secondary osteoarthritis may occur after subsidence of the inflammatory process (burned out RA).

6. *Subchondral cysts.* Cysts are *common,* communicate with the synovium, and may be so large (especially in the hips and knees) that they simulate tumor. The cysts generally do not have sclerotic margins.
7. Periostitis and enthesopathy. Do not occur in RA.
8. *Ankylosis: Very rare* in RA, limited to the carpals and tarsals.
9. Ligamentous abnormality: Common, with tendon ruptures, laxity, and contractures leading to deformities and altered function.

G. Most common joint distribution:
  1. Hand: Some of the earliest findings are in the *MCP* (especially the radial side of the metacarpal head) and *proximal interphalangeal (PIP)* joints (Fig 2–1); distal interphalangeal joints (DIPs) are spared early in the disease. Characteristic deformities are the following:
     a. MCP ulnar deviation and volar subluxation, often with pressure erosions.
     b. Swan-neck: PIP hyperextension and DIP hyperflexion.
     c. Boutonnière: PIP hyperflexion and DIP hyperextension.
     d. Hitchiker's thumb: MCP flexion and interphalangeal (IP) extension.
  2. Wrist: Early erosions (see Fig 2–1) are found in the *distal radioulnar* joint, *ulnar styloid, radial styloid,* waist of *scaphoid, triquetrum,* and *pisiform* (the latter two seen best on the oblique or ball-catcher view; Fig 2–2). Later changes involve the intercarpal and intermetacarpal joints; ulnar "capping" is also seen late in the disease; typical wrist deformities include (see Fig 2–1) *ulnar translocation* (more than 50% of the lunate articulates with the ulna; often associated with radial deviation of the hand), *scapholunate dissociation, dorsi- and palmar*

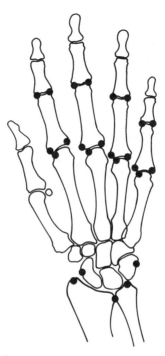

FIG 2–1.
PA view of the hand in RA with characteristic distribution of erosions and ulnar translocation of the carpus.

> *flexion carpal instability patterns,* and *distal radioulnar dissociation.*
3. Elbow: The entire articulation is involved with a positive fat-pad sign, indicating effusion, and erosions of the distal humerus, radial head, and coronoid.

4. Shoulder (Fig 2–3).
   a. Early changes involve *lysis of the distal clavicle,* erosion at the coracoclavicular ligament insertion, and *marginal humeral head erosion* (adjacent to the greater tuberosity, at the insertion of the capsule on the anatomic neck).
   b. Later, secondary signs of *rotator cuff tear* are seen: humeral head elevation and

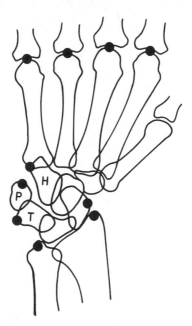

FIG 2–2.
Ball-catcher (oblique) view of hand shows erosions on MC heads, triquetrum, and pisiform to advantage.

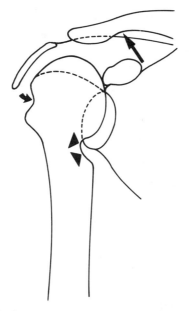

FIG 2–3.
RA of the shoulder with lysis of the distal clavicle, erosion at the coracoclavicular ligament insertion *(long arrow)*, marginal humeral head erosion *(short arrow)*, elevation of the humeral head secondary to rotator cuff tear, and mechanical erosion of the surgical neck of the humerus *(arrowheads)*.

concavity on the under surface of the acromion. With humeral head elevation, *mechanical erosion of the medial surgical neck of the humerus* against the inferior glenoid occurs, occasionally resulting in a pathologic surgical neck fracture.

5. The sternomanubrial and sternoclavicular joints frequently have erosions but are imaged infrequently.
6. Feet.
   a. *MTPs* very commonly have erosive changes (especially on the medial side of the metatarsal (MT) heads), often before wrist and MCP changes appear.
   b. Associated deformities are lateral deviation at the MTPs, *hammer toes* (flexion of PIPs or DIPs), and *cock-up* deformities (hyperextended MTPs).
   c. Intertarsal erosions occur late in the disease.
   d. *Calcaneal spurs* (plantar and at Achilles tendon insertion) may be seen, often with erosions. *Retrocalcaneal bursitis* may obliterate the normal pre-Achilles fat triangle.
7. Ankle: Less commonly involved.
8. Knee (Fig 2–4): Soft tissues demonstrate *suprapatellar effusions* and *popliteal synovial cysts* (Baker's cysts), which may be very large and present as mass lesions. *All three compartments* (medial, lateral, patellofemoral) *demonstrate symmetric cartilage loss, erosion, and subchondral cyst formation.* Deformity most commonly is in *valgus* position. The distal femoral shaft often has an anterior mechanical erosion from patellar pressure.
9. Hip: *Concentric decrease in joint space* and resultant *protrusio* deformity; secondary osteoarthritis is common in the hips, and the two processes may make the diagnosis difficult.
10. Sacroiliac (SI) joints: Involvement by RA is infrequent, mild, and unilateral or asymmetric.
11. Spine: Cervical region is much more commonly involved than thoracic or lumbar.

a. *Atlantoaxial subluxation* (Fig 2–5, A):
Atlantoaxial distance greater than 2.5 mm,
secondary to transverse ligament laxity.
b. *Atlantoaxial impaction* (Fig 2–5, B) due to
C_{1-2} facet erosion is perhaps best detected
by observation of the relationship of the
anterior arch of the atlas with the
odontoid process. On the lateral film, the

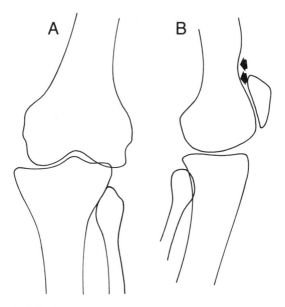

FIG 2–4.
The knee in RA, showing three-compartment disease,
valgus deformity, and mechanical erosion on anterior
femoral shaft *(arrows).*

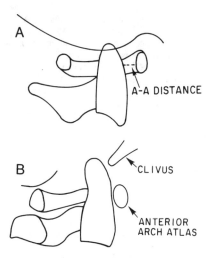

FIG 2–5.
A, RA with atlantoaxial subluxation. **B,** RA with atlantoaxial impaction. Note that with impaction the anterior
arch of the atlas articulates with the body of C$_2$ rather
than with the odontoid (as in **A**). Bilateral facet erosion
and collapse at C$_{1-2}$ allows this impaction.

atlas usually articulates with the cranial
portion of the odontoid but, with
impaction, articulates with the body of C$_2$;
neurologic symptoms more often are
associated with atlantoaxial impaction than
subluxation.
c. *Odontoid erosion.*
d. *Unilateral facet erosion* and collapse at C$_{1-2}$
may result in torticollis and ipsilateral
facial pain.
e. Erosions of the facets and joints of
Lushka.

     f. *Discitis* at several levels thought to be due to a combination of osteoporosis and posterior ligament laxity.

     g. *Mechanical spinous process erosion.*

H. Bilateral symmetry: *Symmetry is generally maintained,* especially regarding groups of joints rather than individual ones. *Exception is patients with neurologic deficits:* the affected side is protected from rheumatoid changes.

I. Other features:
   1. *"Robust RA":* A variety featuring large subchondral cysts and normal bone density, generally seen in patients who maintain their normal activity.
   2. *Felty's syndrome: RA, splenomegaly, and leukopenia.*
   3. *Sjögren's syndrome:* Keratoconjunctivitis sicca, xerostomia, connective tissue disease (often RA).

J. Differential diagnoses:
   1. Psoriatic arthritis: One form resembles the hand and wrist changes of RA, but DIP distribution tends to predominate.
   2. Reiter's disease: Retrocalcaneal bursitis may appear identical to that of RA, but other distribution is usually characteristic (SI joint, feet more prominent than hand changes).
   3. SLE: Same deformities but rarely erosive.
   4. Septic joint: Monostotic, acute onset.
   5. Spondyloarthropathy of hemodialysis may have similar discovertebral junction abnormalities, but C_{1-2} and the facet joints are usually normal, distinguishing this from RA.

K. Survey films for diagnosis of early disease:
   1. Hands: Posteroanterior (PA) and ball-catcher view.
   2. Feet: Anteroposterior (AP).
   3. Lateral cervical spine.
   4. Other symptomatic joints.

Juvenile Chronic Arthritis (Including Juvenile Rheumatoid Arthritis—JRA)

> *Key Concepts:* Group of diseases with systemic symptoms and arthropathy; bones are small and gracile; osteoporotic; symmetric distribution; differs from RA in that fusion and periostitis may be present; predilection for large joints, especially hips, elbows, and knees; metaphyseal flaring due to hyperemia.

A–E. Definition: A group of related diseases of unknown etiology arising in childhood with the following symptom complexes:[1]

1. *Still's disease:* 20% of JRA; *acute systemic disease* occurring in children under 5 years of age. Males and females are affected equally. They present with high fever, anemia, polymorphonuclear leukocytosis, hepatosplenomegaly, lymphadenopathy, and polyarthritis; do not have iridocyclitis. *Radiographic findings are often mild and may not demonstrate erosions;* however, 25% have chronic and destructive arthritis. Another entity, *adult Still's disease,*[2,3] occurs in patients over 18 years of age who present with systemic manifestations and are persistently RF negative; radiographic changes may be identical to adult RA or may be similar to psoriatic arthritis, with predominantly DIP erosion and fusion. Usually, few joints are involved.

2. *Pauciarticular disease:* Most common type (40% of JRA); found predominantly in *young girls.* Involvement of *one to three joints* is rarely severe (usually large joints—knee, ankle, elbow). Chronic *iridocyclitis* occurs in 25%; RF negative and often antinuclear antibody (ANA) positive.

3. *Seronegative polyarticular disease:* 25% of JRA; synovitis with adult type of symmetric and widespread distribution (both large and small joints); female preponderance; occurs at any age; rheumatoid factor negative; no other systemic complaints.

4. *Seropositive polyarticular disease (juvenile-onset adult RA):* 5% of JRA; *polyarticular changes typical of RA;* multiple joints involved, usually sparing the DIPs; severe erosive changes; generally found in *teenage girls;* most *become RF positive;* like adult RA, may have subcutaneous nodules.

5. *Juvenile ankylosing spondylitis:* Findings typical of AS (SI joint and lumbar spine disease) found in adolescent males; RF negative and HLA B27 positive; *often misdiagnosed as JRA* simply because AS may not be considered in a patient so young and because other joints may be involved prior to the SI joints. Finally, SI joint abnormalities may be underdiagnosed because adolescent SI joints are normally wide, with somewhat indistinct cortices.

6. *Psoriatic and inflammatory bowel disease arthropathies* may also be seen in juvenile patients and should be considered in unusual cases of JRA.

F. General radiographic description of the arthropathy of JRA (similar in entities 1–4, above):

1. Soft tissues: May have muscle wasting with severe disease. Periarticular swelling is fusiform.

2. Abnormal calcifications: Rarely are found in juxta-articular locations.

3. *Bone density: Osteoporosis* due to hyperemia and disuse; also, metaphyseal lucent lines (similar to those found in leukemia) may be found during disease activity.

4. *Cartilage destruction: Often a later manifestation than in adult RA.*
5. Erosive vs. productive changes: *Erosive changes occur, generally as a late manifestation.*
6. Subchondral cysts: Rarely present.
7. Periostitis: *Periosteal reaction may be seen early in JRA* (and is virtually never seen in adult RA). Differential considerations of isolated periosteal reaction in a child include osteomyelitis, psoriatic arthritis, and dactylitis (sickle cell infarcts or tuberculosis); Enthesopathy is not seen.
8. Ankylosis: *Fusion is much more common in JRA than adult RA.* Carpus and IP joints are most commonly involved.
9. Ligamentous abnormalities: *Joint contractures are common.*

G. Joints most commonly affected in JRA: Generally, there is a *predilection for large joints* rather than small ones.
   1. Hand and wrist: MCP and PIP involvement similar to adult RA. In the wrist the *radiocarpal joint may be spared and the midcarpal joint, involved,* especially the pericapitate region,[3] distinguishing JRA from adult RA. Adult Still's disease is similar in this aspect. In JRA, *ankylosis is very common in the hand.*
   2. *Elbow:* Commonly and distinctively involved: there is effusion, often *enlargement of the trochlear notch* due to extensive pannus, as well as *radial head enlargement* due to overgrowth, and uniform cartilage loss and destructive change.
   3. Shoulder: Enlarged glenoid and humeral head with late erosion.
   4. Foot and ankle: Commonly involved with MTPs and tarsal joints.
   5. *Knee:* Distinctive involvement, with *effusion,*

widened intercondylar notch from pannus formation, *metaphyseal and epiphyseal flaring and overgrowth, patellar squaring,* and uniform cartilage loss and destructive change.

6. *Hip: Common* distinctive involvement, with *femoral head enlargement, short neck with coxa valga, and* significant *protrusio acetabuli.* The iliac wings are often hypoplastic and femoral shaft, very gracile. This constellation of abnormal shapes and sizes makes prosthesis placement difficult.

7. SI joints: Rare asymmetric involvement; if present, should consider juvenile ankylosing spondylitis. Evaluation of SI joints in the adolescent is particularly difficult because of normal widening.

8. Spine: *Cervical spine* may be commonly involved. *Facet joint erosions and ankylosis* are most common in the upper cervical region. The facet ankylosis is thought to protect JRA patients from developing the discovertebral junction abnormalities seen so often in adult RA. With ankylosis there is often *vertebral body hypoplasia* (both in height and AP diameter). *Atlantoaxial subluxation* (greater than 3.5 or 4 mm in children) and *odontoid erosions* are also prominent findings.

9. Temporomandibular joint (TMJ) erosive changes and micrognathia are relatively common.

H. Bilateral symmetry may be present but is not as reliable as in adult RA.

I. Other features: One very distinctive feature of JRA is *growth abnormalities.* With hyperemia, there is *overgrowth of epiphyses* leading to "ballooning" of joints. *Squaring of carpals, tarsals, and patella* is also seen; however, the hyperemia also leads to *advanced skeletal maturation and premature fusion, resulting in limb length discrepancies and overall shortening of limbs.*

J. Differential diagnoses:
1. Psoriatic arthritis can occur in juveniles and may be indistinguishable.
2. AS may also occur in juveniles, usually males; it affects the same regions as adult AS (SI joints, lumbar spine, hips) and is HLA B27 positive.
3. The radiographic appearance of the knee may be indistinguishable from that in hemophilia or TB arthritis.
4. Eight to 10-year-olds often have irregularities on the articular surfaces of the knees; this normal variant may be misdiagnosed as early erosion of JRA. As described, the constellation of findings—growth deformities, effusion, erosive changes, large joint involvement, periostitis, and ankylosis—in a child usually make the diagnosis of JRA relatively easy.

K. Survey films for diagnosis of early disease:
1. Hands and wrists PA.
2. Lateral cervical spine.
3. Knees.
4. Hips.
5. Other symptomatic areas, especially elbows or ankles.

Ankylosing Spondylitis (AS) and Inflammatory Bowel Disease (IBD) Spondylitis

Key Concepts: Bilateral, often symmetric sacro-iliitis; syndesmophytes and fusion of facet joints of spine, usually without skip areas. Large joints (hips, shoulders) may also be involved.

A. Definition: AS is the most common *seronegative spondyloarthropathy,* of unknown etiology, involving primarily the *axial skeleton and large proximal joints.*

B. Epidemiology:
   1. Sex: Much more common in *males* than females (ratio of 4–10:1, depending on series).
   2. Age: Onset usually *between 15 and 35 years* of age.
   3. *Familial,* but mode of inheritance is unclear.

C. Clinical signs:
   1. *Low back pain,* aggravated by a supine resting position.
   2. *Spine stiffness,* with later postural changes (*increased thoracic kyphosis* and decreased lumbar lordosis).
   3. *Limited chest expansion* (1 inch or less).

D. Laboratory tests:
   1. *HLA B27 positive in greater than 90%* (6% to 8% of the normal population are positive, as are 50% to 80% of patients with Reiter's disease).
   2. RF negative.
   3. ESR increases during disease activity.

E. Extra-articular manifestations:
   1. Iritis.
   2. Heart disease, especially aortic insufficiency.
   3. Pulmonary interstitial disease and fibrosis (especially upper lobes).

F. General radiographic description:
   1. Soft tissues: Generally no change.
   2. Abnormal calcifications: See section G for discussion of syndesmophytes.
   3. *Bone density: Normal* until ankylosis becomes so debilitating that disuse osteoporosis occurs.
   4. Cartilage destruction: Yes.
   5. *Combination of erosive and productive bony changes:* Erosions are smaller and much less prominent than those of RA.

6. Subchondral cysts: Occur, but are not prominent.
7. Periostitis occurs infrequently. *Enthesopathy is common*, especially in the pelvis, calcaneus, and patella.
8. *Ankylosis: Common in SI joints and spine (bodies as well as posterior elements).*
9. Ligamentous abnormalities: Instability not common. Calcification of longitudinal ligaments of the spine may be seen very late in the disease.

G. Joints most commonly affected: The *distribution involving SI joints, spine, and large proximal joints is classic* and crucial in making the diagnosis (Fig 2–6).

1. *SI joints:* Classically the *site of initial involvement;* first changes are loss of cortical definition followed by erosions and joint widening (the findings are most prominent on the iliac side of the joint); later get sclerosis and fuse. The abnormalities may initially be asymmetric but become *bilaterally symmetric* late in the disease process.

2. Thoracolumbar spine: Involvement classically *follows SI abnormalities* and begins at the *thoracolumbar and lumbosacral junctions* and extends contiguously *without skip areas* (skips and asymmetry *may* be seen, but not as commonly as in Reiter's or psoriatic arthropathy). Vertebral involvement *begins with osteitis* (erosive changes at the anterior corners of the vertebral bodies). *"Shiny corners"*—reactive sclerosis at the sites of osteitis—may be seen. The osteitis *leads to* loss of the normal concavity of the anterior vertebral body *(squaring). Syndesmophytes (thin vertical ossifications) form in the annulus fibrosus* at the discovertebral junction (as opposed to bulky horizontal osteophytes that arise from the vertebral body itself, which are seen in

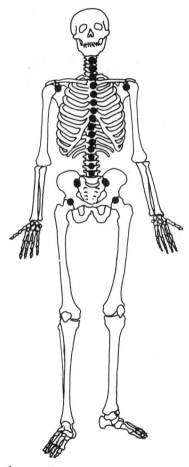

FIG 2–6.
Joints most commonly involved in ankylosing spondylitis.

Reiter's syndrome and psoriatic arthropathy).
Over several segments, *ankylosis* and a
"bamboo" spine occur, with a change in normal
postural alignment. *Fusion of the apophyseal
joints* also occurs: the anterior longitudinal
ligament may eventually ossify; the fused and
osteoporotic spine is vulnerable to fracture
from minor trauma and pseudarthrosis forms
easily. Pseudoarthroses are most commonly
seen at the cervicothoracic and thoracolumbar
junctions; the pseudoarthrosis goes through
the disc space anteriorly, and continues
through the posterior elements, often appearing extremely subtly. *NB: All the spine changes
are best seen on the lateral film.*

3. Cervical spine: Generally involved late, in an
   ascending fashion from the thoracic spine.
   Odontoid erosion and atlantoaxial subluxation
   are seen, but less commonly than in RA.
4. *Hip: Most common appendicular joint involved*
   (up to 50% of AS patients); *concentric joint
   narrowing, mild erosions, protrusio acetabuli, and
   ring osteophytes give appearance of a combination
   of RA and OA.* In a young patient, these
   incongruent findings alone should suggest a
   diagnosis of AS and lead to careful scrutiny
   of the SI joints. Involvement is often bilateral
   but may be asymmetric.
5. *Glenohumeral:* Next most commonly involved
   appendicular joint after the hip; again, a
   combination of erosive and productive
   findings is manifest. *Involvement is often
   bilateral but may be asymmetric.*
6. Symphysis pubis, sternomanubrial, and
   costovertebral joints commonly are involved,
   leading to ankylosis.
7. Knees, ankles, hands, and feet are less
   commonly involved (much less often than in
   Reiter's disease or psoriatic arthritis). Usually
   other findings more typical for AS are

present, making the diagnosis less difficult.

H. *Bilateral symmetry: An important feature late in the disease* but asymmetry may occur early.

I. Other features:

1. AS is uncommon, but certainly occurs, in women. The radiographic findings tend to be neither as severe nor as classic in distribution as in male patients.

2. IBD arthritis[4]:

   a. One group of disease processes occurs from *Salmonella, Shigella,* or *Yersinia.* A self-limited polyarthritis may occur, usually without radiographic findings but occasionally with SI joint abnormalities.

   b. Another group of disease processes may occur with ulcerative colitis, Crohn's disease, or Whipple's disease. Ten to 15% of these patients develop an arthropathy; 50% to 60% of these are peripheral arthropathies (predominantly in the lower extremities, in more unusual joints—knees, ankles, elbows—and with milder osseous changes than in AS). The flares of the peripheral arthropathy correlate with disease activity. Twenty to 30% of patients develop a sacroiliitis identical clinically and radiographically to that of AS; the sacroiliitis does not correlate with IBD activity.

J. Differential diagnoses for SI joint disease of AS:

1. Psoriatic arthritis or Reiter's syndrome: Asymmetry of SI disease is more suggestive of these diseases. The spondylitis pattern is also distinctive, with bulky asymmetric osteophytes and skip regions. Finally, the distribution of peripheral arthritis is helpful. (More acral distribution than hips or shoulders tends to occur in both Reiter's and psoriatic arthritis.)

2. IBD may not be distinguishable from AS but tends to be less severe.
3. RA has less severe and rarely bilaterally symmetric SI disease.
4. Hyperparathyroidism (HPTH): The subchondral collapse may simulate SI joint arthritis, but generally other findings of HPTH also are present.

K. Suggested survey films for AS/IBD: AP pelvis (looking for SI disease, hip disease, and enthesopathy) and lateral thoracolumbar spine (looking for vertebral body squaring or syndesmophyte formation).

Psoriatic Arthritis

> **Key Concepts:** Predominantly erosive changes involving the carpus, DIP, PIP, but less commonly MCP; sacroiliitis, often asymmetric, is common; bulky asymmetric osteophytes at thoracolumbar junction; arthropathy may antedate skin changes.

A. Definition: An arthropathy occurring in 0.5% to 25% of patients with psoriasis. *Five distinct manifestations* are described by Wright and Mall:
1. *Polyarthritis* (predominantly *DIP*).
2. *Arthritis mutilans* (deforming type).
3. *Symmetric type (resembling RA)*.
4. *Oligoarthritis*.
5. *Spondyloarthropathy* (occurring in 30% to 50% of patients with psoriatic arthritis).

B. Epidemiology:
1. Age: Generally young adults.
2. Sex: Affects females and males equally.
3. Psoriatic skin disease is usually present prior to the arthropathy, but the *arthropathy may antedate skin findings in 20% of cases.*

C. Clinical signs:
   1. Soft tissue swelling, especially in the *small joints of the hands and feet* may involve an entire digit (*sausage digit*). There is pain and reduced range of motion.
   2. Low back pain.
   3. Nail changes (thickening, pitting, or discoloration) are very common and are highly correlated with severity of the arthropathy.
D. Laboratory tests:
   1. Negative RF.
   2. Elevated ESR.
   3. *HLA B27 positive in 25% to 60%.*
E. Extra-articular manifestations: Psoriatic skin and nail changes.
F. General radiographic description of the arthropathy:
   1. Soft tissue: Swelling at involved joints, either fusiform or involving the entire digit (*sausage digit*).
   2. Abnormal calcifications: May develop an *ivory phalanx*—reactive sclerosis of the tuft.
   3. Bone density: Early juxta-articular osteoporosis may be seen, but the *density is generally normal* (an important distinguishing characteristic from RA).
   4. Cartilage destruction: Joint occasionally is widened, but generally narrowing occurs.
   5. Erosive vs. productive change:
      a. *Erosions begin marginally*, as in RA, but progress to severe subchondral erosions, occasionally resulting in a *pencil in cup deformity* (characteristic of, but not pathognomonic for, psoriatic arthritis).
      b. Productive changes are also seen, usually in the form of excrescences at and around the joint.
   6. Subchondral cysts: Not commonly seen.

7. Periosteal reaction: *Periosteal reaction* is often seen in hand and foot phalanges with the sausage digit pattern of soft tissue swelling. This periostitis may help differentiate psoriatic arthritis from RA. An *enthesopathy* similar to that seen in AS or Reiter's is also common.

8. *Ankylosis:* Common, especially in the *hands and feet* (again, helping to differentiate this disease from RA).

9. Ligamentous abnormality: Not a prominent finding; however, phalanges with severe pencil in cup deformities may telescope.

G. Joints most commonly affected (Fig 2–7): *The characteristic distribution is the small joints of the hands and feet, with or without a spondyloarthropathy.*

   1. *Hand:* Tuft resorption and DIP erosive disease are usually seen earlier and involvement may be more severe than PIP or MCP joints. This pattern helps differentiate psoriatic arthritis from RA. Also, any compartment of the wrist may be involved, but generally only after DIP abnormalities occur. *Asymmetry is far more common than in RA.*

   2. *Foot:* IP and MTP erosive disease are common. *Ivory tufts* are described. A *retrocalcaneal bursitis* may be seen (as in RA or Reiter's syndrome), along with erosions at the sites of the Achilles tendon and plantar aponeurosis insertions.

   3. Larger joint (ankle, knee, hip, shoulder) involvement is *much less common. If large joints are involved, the distal small joints are almost invariably involved as well.*

   4. *SI joints: Common* site of involvement (30% to 50%), with erosions, widening, sclerosis, and, eventually, fusion. The disease is usually bilateral, and symmetry is common, though perhaps less common than in AS.

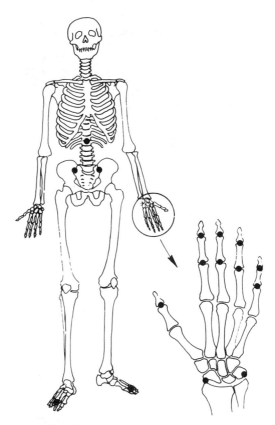

FIG 2–7.
Joints most commonly affected in psoriatic arthritis.

5. *Spine:* Earliest involvement is usually in the region of the *thoracolumbar junction, with the formation of large bulky asymmetric osteophytes*

(the same pattern as in Reiter's, but differing from the syndesmophytes of AS). Findings in the cervical spine are much less distinctive, including discitis, atlantoaxial subluxation, and apophyseal joint disease. These findings alone do not distinguish psoriatic arthritis from RA, AS, or Reiter's.

H. Symmetry:
  1. *Lack of symmetry in the small joints is common*, helping to distinguish from RA.
  2. *SI symmetry is common, but less dependably so than in AS.*

J. Differential diagnoses:
  1. Reiter's disease: SI joint and spine disease are indistinguishable from psoriatic arthritis. Foot disease is usually more severe in Reiter's, while hand disease is a less prominent feature.
  2. RA: Occasionally indistinguishable from one type of psoriatic arthritis, but the predominantly distal distribution in psoriatic arthritis usually leads to the correct diagnosis. Adult Still's disease, with its DIP distribution, may be indistinguishable.
  3. AS/IBD: The SI joint disease may be indistinguishable, but the spine disease is clearly different, as is the small joint distribution as opposed to large joint disease seen in AS/IBD.
  4. Erosive osteoarthritis (EOA): DIP erosions are potentially confusing, but EOA usually has abnormalities in the first carpometacarpal joint or the scaphoid-trapezium-trapezoid joints.
  5. "Erosions" of hyperparathyroidism: Should also see other findings of HPTH, especially subperiosteal erosion.

K. Survey films for early diagnosis:
  1. Hands: PA and ball-catcher views.

2. Feet: AP and lateral views.
3. AP pelvis: For SI joints and enthesopathy.
4. AP view of thoracolumbar spine.

Reiter's Disease

> **Key Concepts:** The predominant arthritic pattern includes a spondyloarthropathy and lower extremity erosive disease. The spondyloarthropathy is identical to that of psoriatic arthritis, with asymmetric or symmetric sacroiliitis and bulky thoracolumbar osteophytes. Calcaneal erosive disease and spur formation are particularly prominent.

A. Definition: A syndrome consisting of the *triad of (1) urethritis (cervicitis in females) in 85% of cases, (2) conjunctivitis in 60% of cases, and (3) arthritis.* The syndrome may be incomplete or may include balanitis or keratoderma blennorrhagicum.
B. Epidemiology:
  1. Age: *Young adult.*
  2. Sex: *Males* affected much more common than females.
  3. The arthropathy only rarely precedes the urethritis or conjunctivitis.
C. Clinical signs: Low back pain, polyarticular arthritis *with heel pain predominating,* urethritis, or conjunctivitis.
D. Laboratory tests:
  1. RF negative.
  2. Elevated ESR.
  3. *HLA-B27 positive in 80%.*
E. Extra-articular manifestations: Besides urethritis and conjunctivitis, pulmonary fibrosis, valvular disease, and diarrhea may rarely be seen.
F. General radiographic description of the arthropathy:

1. Soft tissues: Swelling at involved joints may be either fusiform or include the entire digit (*sausage digit,* seen also in psoriatic arthritis).
2. Abnormal calcifications: None.
3. Bone density: May develop osteoporosis around involved joints, but much less consistently than in RA.
4. Cartilage destruction: Decreased joint space.
5. Erosive versus productive change: *Erosions predominate but may be seen in conjunction with mild productive bony changes.*
6. Subchondral cysts: Rare.
7. Periostitis: *Periosteal reaction in phalanges* seen with sausage digits, as in psoriatic arthritis. *Enthesopathy, especially in the pelvis and calcaneus, is common.*
8. *Ankylosis: Occurs in the SI joint but is much less common elsewhere than with psoriatic arthritis.*
9. Ligamentous abnormalities: Uncommon.

G. Joints most commonly affected (Fig 2–8): *The distribution is predominantly distal lower extremity (MTP, calcaneus, ankle, knee), often with a spondyloarthropathy identical to that of psoriatic arthritis.*

   1. Foot: Earliest changes:
      a. *Small joints of the foot (especially MTPs and first IP) are commonly involved, primarily with erosive changes* (similar to RA, but sausage digit and periostitis, as well as normal hands help to differentiate the two).
      b. *Retrocalcaneal bursitis is common,* along with *prominent spur formation* at the Achilles tendon and plantar aponeurosis insertion; erosive changes also are often present.
      c. Erosions are seen in tarsals with more severe disease.
   2. Other lower extremity joints (ankle, knee, hip) may have an erosive and productive

pattern similar to that of AS or psoriatic arthritis. Hip involvement is less common in Reiter's disease than in the other rheumatoid variants.

3. Upper limb involvement: Rare.
4. *SI joint involvement: Common* (10% to 40%), generally *bilateral* and may be *symmetric or asymmetric* (especially early in the disease process). The appearance of SI joint

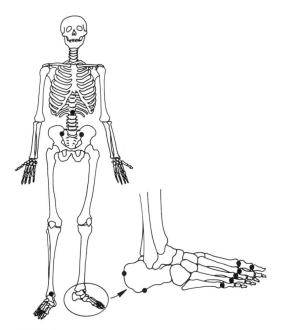

FIG 2–8.
Joints most commonly involved in Reiter's disease.

widening, sclerosis, erosions, followed by fusion is *identical to that seen in psoriatic arthritis.*

5. *Spine: Bulky asymmetric paravertebral osteophytes are often first seen in the thoracolumbar regions, commonly skipping segments.* The appearance is *identical to that of psoriatic arthritis* and is best noted on the AP film, as opposed to AS, where the lateral film best demonstrates the thin vertical syndesmophytes and vertebral body squaring.

H. Symmetry:
   1. *Lack of symmetry* is common in the foot disease, especially early.
   2. *SI joint disease is often symmetric but less dependably than in AS* (especially early).

J. Differential diagnoses:
   1. Psoriatic arthritis: The spondyloarthropathy is identical, the foot disease is identical, but the sites of predominant peripheral involvement usually differs, with hand disease predominating in psoriatic arthritis and foot disease predominating in Reiter's disease.
   2. AS/IBD: The sacroiliac joint disease is similar, but spine disease is usually quite distinctive. The appendicular distribution is significantly different, with large joint involvement in AS/IBD but primarily foot involvement in Reiter's.
   3. RA: May be confused with Reiter's if there is primarily MTP or calcaneal involvement early in the disease; rheumatoid factor and symmetric hand abnormalities usually distinguish the two.

K. Survey films for early diagnosis:
   1. AP and lateral feet.
   2. AP pelvis looking for SI joint disease.
   3. AP thoracolumbar spine.

II. CONNECTIVE TISSUE DISORDERS
Systemic Lupus Erythematosus (SLE)

> ***Key Concepts:*** Generally nonerosive deforming arthropathy (but reducible). AVN is very common.

A. Definition: SLE is an immunologic abnormality that produces ANAs that in turn cause severe and widely varied tissue injury. *The musculoskeletal system is most commonly involved with a polyarthritis that is generally nonerosive but may be deforming. There are frequent remissions and exacerbations with severe symptoms, but these are accompanied by remarkably few radiographic abnormalities.*

B. Epidemiology:
   1. Sex: *Female* much more common than males (5 to 10:1).
   2. Race: More *blacks* than Caucasians are affected.
   3. Age: *Young adults* (generally under age 40).

C. Clinical signs:
   1. *Polyarthritis in 90% of cases; myositis.*
   2. *Typical skin rash.*
   3. *Constitutional signs:* Malaise, weakness.
   4. Multiple system disease (see E, below).

D. Laboratory tests:
   1. Lupus erythematosus (LE) cell prep positive.
   2. ANA positive.
   3. *RF may yield a false positive.*

E. Extra-articular manifestations:
   1. Myositis.
   2. Rash.
   3. Various neurologic abnormalities.
   4. Pulmonary vasculitis, fibrosis, and effusions.
   5. Pericarditis, cardiomyopathy.

   6. Nephritis.
   7. Steroid-induced stress fractures.
F. General radiographic description:
   1. Soft tissue: Symmetric swelling around joints, not chronic.
   2. Calcifications: Subcutaneous calcifications, generally *in the lower extremities, uncommon* (less than 10%).
   3. *Bone density:* Usually *normal,* but periarticular demineralization occasionally is seen.
   4. *Cartilage destruction: Rare.*
   5. *Erosions or productive change: So rare* that if erosions are seen, either early RA or a mixed connective tissue disease or overlap disease is considered.
   6. Subchondral cysts: Rare.
   7. Periostitis or enthesopathy: Not present.
   8. Ankylosis: Does not occur.
   9. *Ligamentous abnormality: The classic descriptor of SLE is nonerosive deformity from ligamentous laxity.* This is seen in only approximately 10% *of cases.* The deformities are most commonly seen in the *hand,* with *ulnar deviation of the MCPs,* subluxation of the first carpometacarpal joint, and *variable flexion or extension deformities of the IP joints.* The deformities are *reducible,* so they are more prominent on the oblique ball-catcher view of the hand than on the PA view, where the hand and fingers are supported by the film cassette.
G. Joints most commonly affected: *Hand, wrist, knee—all characterized either by nonspecific polyarticular swelling or by a nonerosive ligamentous deformity.*
H. Bilateral symmetry: Generally present.
I. Other features: *Remarkably high incidence of AVN* (up to one third of SLE patients, though only 8 percent are symptomatic).[5] *Steroid therapy is felt to be the major etiologic factor, but the disease process*

itself may also predispose to AVN. Femoral head, humeral head, and knee are common but nonspecific sites. The *dome of the talus and hands or feet are sites that are not uncommon in SLE but extremely rare sites for AVN of another etiology;* therefore, AVN in these sites suggests SLE.

J. Differential diagnoses:
  1. *RA:* The symmetric polyarthritis, deformities, and false-positive RF may be confusing in SLE, but the lack of erosions should make one suspect SLE rather than RA.
  2. *Jaccoud's arthropathy:* Identical nonerosive deformities; history of rheumatic fever makes the diagnosis.
  3. If the radiographic findings are restricted to AVN, all the other causes of AVN might be considered.

K. Survey films:
  1. PA and ball-catcher views of hands.
  2. If painful, the hips, shoulders, knees, or ankles may be studied for signs of AVN.

Progressive Systemic Sclerosis: Scleroderma

> **Key Concepts:** Acral soft tissue atrophy, acro-osteolysis, soft tissue calcification; erosions occur, but are not a prominent feature.

A. Definition: A condition of unknown etiology that causes small-vessel disease and fibrosis in several organ systems. Scleroderma is the cutaneous manifestation of the disease.

B. Epidemiology:
  1. Sex: Affects *females* more often than males (3:1).
  2. Age: Most often diagnosed in the third to fifth decades.

C. Clinical signs:
  1. May present with Raynaud's phenomenon.

    2. Skin changes on the hands, feet, or face.
    3. Distal joint pain and stiffness.
    4. Dysphagia.
    5. Proximal myopathy.
  D. Laboratory tests: *Not specific.*
    1. ESR increased in 70%.
    2. ANA increased in up to 95%.
    3. RF positive in up to 40%.
  E. Extra-articular manifestations:
    1. Skin changes: Edema, leading to thickening and fibrosis, eventually becoming taut, shiny, and atrophic with progressive distal tapering.
    2. GI: Esophageal atrophy and fibrosis, leading to dysmotility (air-fluid levels may be seen within the esophagus); pseudosacculations seen in the colon.
    3. Pulmonary fibrosis.
    4. Renal fibrosis.
    5. Cardiac: Pericarditis and myocarditis.
  F. General radiographic description:
    1. Soft tissue: Extremely common; *acral tapering of digits* (78% in one series).[6]
    2. *Calcification: A prominent feature, with extensive subcutaneous, extra-articular, and occasionally intra-articular calcification;* punctate calcification in the terminal phalanx may also be seen. (Calcification seen in 25% of Bassett's series[6]).
    3. Bone density: Generally normal, but periarticular osteoporosis may be seen.
    4. Cartilage destruction: Joint space narrowing occasionally is seen.
    5. Erosive vs. productive change:
      a. The issues regarding true erosive change in scleroderma are difficult to resolve since, clinically, 25% to 50% of cases of scleroderma look like rheumatoid arthritis as well (often even with a positive RF). Thus many patients have concurrent RA or overlap syndromes that explain any observed erosive changes.

b. In addition, some scleroderma patients who are RF negative also have erosions, most commonly in the PIP and DIP joints.

c. *Overall, joint abnormalities eventually occur in nearly 50% of scleroderma patients, predominantly erosions, but there may also be mild productive changes.*

6. Subchondral cysts: None.

7. Periostitis or enthesopathy: None.

8. Ankylosis: May occur.

9. Ligamentous abnormality: *Flexion contractures are common*, especially in the hands, wrists, and elbows.

G. Joints most commonly affected: PIPs and DIPs, sparing the MCPs and wrists.

H. Bilateral symmetry: Often present.

I. Other features: *Resorption of bone is extremely common (seen in up to 80%):*

1. *Tufts: Acro-osteolysis*, initially on the palmar aspect.

2. *Severe resorption at the first carpometacarpal (CMC) joint* (trapezium and base of first metacarpal) *with radial subluxation of the first MC is very distinctive and almost pathognomonic of scleroderma.*

3. Resorption at the angle of the mandible is seen with facial skin changes.

4. Posterior resorption of ribs 3 through 6 (related to intercostal muscle atrophy).

J. Differential diagnoses:

1. *Differential for acro-osteolysis* includes hyperparathyroidism and thermal injury, among many others; the soft tissue tapering and frequent calcifications help make the diagnosis of scleroderma.

2. *Differential for the erosive changes* includes RA (the sparing of the wrists and MCPs helps differentiate it from RA), and psoriatic arthritis (again, soft tissue calcifications, if present, help differentiate scleroderma).

3. *Differential of soft tissue calcification* is very extensive, including dermatomyositis, hyperparathyroidism, hypoparathyroidism, tumeral calcinosis, metabolic abnormalities involving vitamin D, and HA crystal deposition disease.

K. Suggested survey films: PA and lateral hands.

Polymyositis/Dermatomyositis

> **Key Concepts:** Soft tissue calcifications, subcutaneous or sheetlike in fascial planes; usually no joint abnormalities despite arthralgias.

A. Definition: A disease of unknown etiology that produces *inflammation and muscle degeneration*. In polymyositis the symptoms of proximal muscle weakness and arthralgias predominate; with dermatomyositis, a typical diffuse rash is an additional finding.

B. Epidemiology:
   1. Sex: *Females affected more than males.*
   2. Age: Third through fifth decades; dermatomyositis may also be seen in children, associated with severe systemic symptoms.

C. Clinical signs:
   1. *Muscle weakness*, tenderness, and eventually contracture with atrophy (50%).
   2. *Rash* (50%).
   3. *Raynaud's phenomenon* (33%).
   4. *Arthralgias* (20% to 50%).

D. Laboratory tests: Elevated muscle enzymes during active disease.

E. Extra-articular manifestations:
   1. Pulmonary fibrosis.
   2. Pericarditis.
   3. Abdominal pain, dysphagia.

F. General radiographic description:

1. *Soft tissue: Muscle edema early, followed by atrophy and calcification.*
2. Calcification:
   a. *Subcutaneous calcification* is most commonly seen and is nonspecific.
   b. *Sheetlike calcification along fascial or muscle planes* is less common but nearly pathognomonic for the disease; classically, it is seen in the proximal large muscles.
   c. Periarticular calcification may also occur.
3. Bone density: Transient periarticular osteoporosis occasionally.
4. Cartilage destruction: Not seen.
5. Erosive change: If present, generally indicates an overlap syndrome; sporadic reports of erosions in dermatomyositis exist.
6. Subchondral cysts: None.
7. Periostitis: None.
8. Ankylosis: None.
9. Ligamentous abnormalities: May get flexion deformities.

G. Joints most commonly affected with arthralgias: Hands, wrists, knees; *radiographic abnormalities in joints are rare.*

H. Bilateral symmetry: Common.

I. Other features: When dermatomyositis develops in older males, it may be associated with malignancy.

J. Differential diagnoses: If calcification is not sheetlike: scleroderma, SLE, overlap syndrome, HPTH.

K. Survey films: Symptomatic sites.

Amyloidosis

> **Key Concepts:** Nodular synovitis that is very bulky. Erosions are better marginated than in RA; wrist, elbow, shoulder most commonly involved.

A. Definition: An infiltrative disorder that may be either *primary or secondary* (associated with other disease processes such as multiple myeloma, rheumatoid arthritis, familial Mediterranean fever, chronic infection, spondyloarthropathy, and connective tissue disorders such as SLE, scleroderma, and dermatomyositis). *Five to 13% of patients with amyloid have bone or joint involvement; this may consist of deposition in bone, synovium, and surrounding soft tissues.*

B. Epidemiology:
   1. Sex: *Males* affected more commonly than females.
   2. Age: Fourth through 8th decade.

C. Clinical signs of the arthropathy: Pain, stiffness, and soft tissue swelling; joint contractures and carpal tunnel syndrome may occur.

D. Laboratory tests: Biopsy often required to confirm the diagnosis.

E. Extra-articular manifestations:
   1. Kidney infiltration: Most common site of involvement; causes the greatest morbidity.
   2. Organomegaly.
   3. Pericardial and myocardial infiltration may lead to cardiac failure.
   4. Pulmonary septal infiltration.
   5. GI tract may be involved from the tongue to anus. Submucosal thickening and decreased peristalsis are observed.

F. Radiographic description of the arthropathy:
   1. *Soft tissue: Bulky nodules,* especially about *wrists, elbows, and shoulders (shoulder pad sign:* bulky nodules superimposed on atrophic musculature).
   2. Abnormal calcifications: None.
   3. Bone density: May be diffusely or focally osteoporotic.
   4. Cartilage destruction: *Joint space may actually widen* due to infiltration.

5. Erosions: *Well-marginated erosions* occur with intra-articular disease.
6. Subchondral cysts: May occur.
7. Periostitis or enthesopathy: None.
8. Ankylosis: None.
9. Ligamentous abnormality: *Joint contractures* occur.

G. Joints most commonly affected: Wrist, elbow, shoulder; knees and hips less commonly.

H. Bilateral symmetry: Common.

J. *Differential diagnosis* of the arthropathy: *RA is the prime consideration, but the well-defined erosions, preservation of joint space, and nodular soft tissue may help to differentiate amyloidosis from RA.*

III. RHEUMATIC FEVER

> **Key Concepts:** Nonerosive but deforming arthropathy (especially at MCP joints) occurs uncommonly.

A. Definition: Rheumatic fever is a syndrome that follows a group A β-hemolytic streptococcal infection (usually of the throat) and produces fever, various systemic symptoms, valvular heart disease, a polyarthritis and, in a few cases, Jaccoud's arthropathy (a deforming nonerosive arthropathy).

B. Epidemiology: Not useful.

C. Clinical signs: The more common *polyarthritis* presents with joint pain, sometimes accompanied by signs of inflammation and swelling. *Large joints* (knee, ankle) are involved most commonly, and the pattern is often *migratory*. Radiographs are normal except for occasional nonspecific soft tissue swelling and juxta-articular osteoporosis resulting from the local inflammation. The arthritis is self-limited, disappearing after a few weeks. The much less common *Jaccoud's arthropathy* generally occurs following multiple episodes of polyarthritis and is *clinically asymptomatic.*

D. Laboratory tests: RF is negative.

E. Extra-articular manifestations: Fever; valvular heart disease.

F. *General radiographic description of Jaccoud's arthropathy (nonerosive but deforming):*
   1. Soft tissue: Normal.
   2. Abnormal calcifications: None.
   3. Bone density: Usually normal, though juxta-articular osteoporosis may occur.
   4. *Cartilage destruction: Does not occur until very late and then is secondary to mechanical wear due to subluxation.*
   5. *Erosions: The disease is nonerosive until very late, when hook erosions on the radial aspect of the metacarpal heads, away from the articular margin, may occur secondary to mechanical pressure.*
   6. Subchondral cysts: Rare.
   7. Periostitis or enthesopathy: None.
   8. Ankylosis: None.
   9. *Deformity: The primary feature of the arthropathy is a reversible deformity (ulnar deviation and flexion of the MCPs and fibular deviation and flexion of the MTPs); occurs secondary to capsular and tendon fibrosis.*

G. Joints most commonly affected: By *polyarthritis: knees and ankles;* by *Jaccoud's arthritis: MCPs and MTPs.*

H. Bilateral symmetry: Often occurs, but not prominently.

I. Other features: None.

J. *Differential diagnoses* for Jaccoud's arthropathy:
   1. *SLE:* May be indistinguishable radiographically in the hands or feet.
   2. *RA:* The deformity is identical, but RA rarely causes the deformity without accompanying erosions and cartilage damage. When

erosions occur in Jaccoud's they are not the typical marginal erosions seen in RA.

3. Ehlers-Danlos syndrome: May rarely appear similar.

K. Suggested survey films: PA and ball-catcher (oblique) views of hands: The reversibility will be demonstrated since the hands appear nearly normal on the PA (where they are supported by the cassette) but the deformities are marked on the unsupported oblique views.

IV. OSTEOARTHRITIS (OA)

> *Key Concepts:* Most common locations: DIPs, first CMC, scaphoid-trapezium-trapezoid, lower cervical spine, lumbar facets, hips, knee, first MTP. Normal bone density, focal (weight-bearing) loss of cartilage, sclerosis, osteophyte formation—productive changes are prominent.

A. Definition: Degenerative joint disease stimulated by one or a combination of the following factors (NB: In most cases, osteoarthritis is believed to be secondary, with an underlying cause which usually can be found):
   1. Abnormal mechanical forces on the joint (joint deformity, obesity, occupational stresses).
   2. Normal forces on abnormal cartilage (due to a pre-existing arthritis such as RA, loose bodies in the joint, osteochondral fracture, or a meniscal abnormality in the knee).
   3. Collapse of subchondral bone (due to osteoporosis, avascular necrosis, or hyperparathyroidism).
B. Epidemiology:
   1. The most common arthritis.
   2. Sex: Males and females are affected equally

  though the disease often presents earlier in
  males.
  3. Age: Incidence increases with age.
C. Clinical signs:
  1. Pain (on bearing weight, relieved with rest).
  2. Limited range of motion.
  3. Crepitus.
  4. Subluxation, most commonly genu varus.
  5. Heberden's (DIP) and Bouchard's (PIP)
     nodes.
  6. The radiographic severity of disease does not
     always correlate strongly with amount of
     pain.
D. Pertinent laboratory tests: None.
E. Extra-articular manifestations: None.
F. General radiographic description:
  1. Soft tissue alterations:
     a. *Heberden's (DIP) and Bouchard (PIP) nodes*
        are actually the osteophytes formed at
        these sites.
     b. Effusions occur but are unusual.
  2. Abnormal calcifications:
     a. May develop *chondrocalcinosis*.
     b. May develop HA depositions in
        periarticular sites.
     c. *Loose bodies* may form from synovial
        metaplasia or fractured osteophytes.
  3. *Bone density:* Normal.
  4. *Cartilage destruction:* Always present and tends
     to be focal, predictably in the primary *weight-
     bearing portion* of the joint.
  5. Erosive versus productive bony change: *No
     erosions.* Three types of *productive change* are
     commonly seen:
     a. *Osteophytes* may be intra-articular but occur
        primarily in *non–weight-bearing sites*, due to
        capsular or ligamentous traction.
     b. *Subchondral sclerosis* due to vascular
        invasion after abnormal mechanical forces

and deposition of new bone in a reparative attempt.

   c. *Cortical buttressing,* another reparative attempt in response to abnormal mechanical forces, is seen primarily in the *medial and lateral aspects of the femoral neck.*

6. *Subchondral cysts:* Common; microfractures in the subchondral bone and synovial fluid pressure probably combine to form cysts, but many do not communicate with the joint. Some "cysts" are areas of fibrocartilaginous metaplasia. The cysts tend to occur in *weight-bearing* areas and generally have a *sclerotic margin* (unlike RA erosions). *Eggar's cyst* is found in the weight-bearing portion of the acetabulum and may be the first sign of hip OA.

7. *Enthesopathy* is *common,* especially on the anterior aspect of the *patella* and the *pelvic and hip apophyses.* It is not distinguishable from that seen in ankylosing spondylitis and other rheumatoid variants.

8. Ankylosis: Rare in absence of trauma; in erosive OA the DIP joints occasionally may fuse.

9. *Ligamentous abnormality: Commonly seen secondary to the underlying joint OA;* the focal cartilage loss leads to joint deformity, which in turn promotes ligamentous contractions and laxity. These deformities lead to instability which in turn promotes further arthropathy.

G. Joints most commonly affected:
   1. Hand:
      a. *Multiple IP joints* are often involved with uniform cartilage narrowing, subchondral sclerosis, and osteophyte formation.
      b. Generally, the IP joints are not painful and function is retained.

    c. Occasionally flexion or even radial or ulnar deformity is seen.

    d. Heberden's and Bouchard's nodes represent the underlying osteophytes and not soft tissue swelling.

    e. MCPs are less commonly involved than IPs, and never are involved in the absence of IP joint OA; subchondral cysts may be prominent in MC heads.

    f. Differential diagnosis for IP disease: Rheumatoid variants, such as psoriatic arthritis, that have a combination of erosive and productive changes; erosive OA (see I, below).

    g. Differential diagnosis for MCP disease: Pseudogout and hemochromatosis have prominent hooklike osteophytes and cysts, generally seen in the second and third MCPs and often with chondrocalcinosis seen in the triangular fibrocartilage.

2. Wrist (Fig 2–9):

    a. Typical changes of sclerosis, osteophyte formation, and cartilage loss involve the *first carpometacarpal joint*, often with radial subluxation of the thumb.

    b. The second most common site is the *scaphoid-trapezium-trapezoid* complex.

3. Elbow: Only posttraumatic involvement.

4. Glenohumeral: OA unusual in absence of previous trauma. When present, there are *marginal osteophytes around the glenoid,* as well as *ring osteophytes around the anatomic neck of the humerus,* largest inferiorly. *If there is no history of trauma, pseudogout* should be considered.

    a. A chronic rotator cuff tear (seen as elevation of the humeral head and a sclerotic concave acromion articulating with the humeral head) may lead to OA of the shoulder.

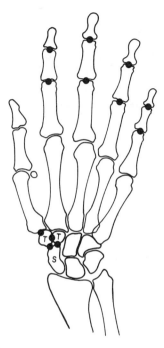

FIG 2–9.
Joints of the hand and wrist most commonly involved
with osteoarthritis.

> b. Shoulder impingement syndrome gives a
> different appearance, with sclerosis of the
> top of the greater tuberosity and
> occasionally a subacromial spur.
> 5. Acromioclavicular joint: Common; typical
> changes of OA.
> 6. SI joint: Subchondral sclerosis and
> osteophytes are commonly seen in two

sites—*anteroinferior to the SI joint and
anterosuperior to the synovial portion of the joint*
(mid to upper third of SI joint as seen on AP
films). Large *osteophytes bridge the joint
anteriorly;* this appearance is easily identified
as an osteophyte on computed tomography
(CT) but may be *misinterpreted as focal blastic
metastases* on plain film.

   a. *Differential diagnoses:*

     (1) Metastases.

     (2) Ankylosing spondylitis.

     (3) Osteitis condensans ileii (a triangular
        sclerotic lesion on the iliac side of the
        inferior SI joint, seen most often in
        multiparous women).

7. Hip: A common site of OA that is often
painful with weight bearing and has a
restricted range of motion. Eggar's cyst of the
acetabulum and calcar buttressing may be
early signs of OA; otherwise expect to see
typical signs of focal cartilage narrowing,
sclerosis, and osteophyte formation (lateral
acetabulum, lateral and especially medial
subcapital region). The hip *migrates superiorly
in 80% of patients,* usually superolaterally,
though sometimes the migration appears to
be superomedial due to cartilage narrowing
and huge inferomedial osteophytes. *Medial
migration with protrusio acetabuli is seen in 20%.*

   a. *Differential diagnoses:*

     (1) If medial migration, look for reasons to
        differentiate it from RA or the
        rheumatoid variants, which should have
        combined erosive and productive
        changes and/or SI joint abnormalities.

     (2) Superior flattening of the femoral head,
        accentuated by the presence of large
        inferomedial osteophytes simulates the
        appearance of *avascular necrosis;* the
        latter disease process has normal

cartilage width, differentiating it from OA, but if secondary OA occurs, the differentiation is not always possible.

(3) If large cysts are present, *pseudogout or PVNS* should be considered.

(4) An early ring osteophyte in the subcapital position may be difficult to differentiate from an *impacted subcapital fracture.*

8. Knee: A common site of involvement, with pain on weight bearing; may involve one, two, or all three compartments (medial, lateral, patellofemoral). If single-compartment disease, it is likely to be medial. Specific compartment involvement should be evaluated since unicompartmental OA may be treated with a unicompartmental prosthesis or with a high tibial osteotomy (closing wedge tibial metaphyseal osteotomy, which transfers much of the weight bearing to the lateral compartment and may lead to regeneration of fibrocartilage in the medial compartment). With medial compartment predominance, one sees a *typical varus knee deformity and lateral subluxation of the tibia.*

a. Osteophytes tend to be marginal in all three compartments, as well as on tibial spines.

b. Enthesopathy of the anterior (nonarticular) surface of the patella at the quadriceps insertion is common and not related to OA.

c. *Differential diagnoses:*

(1) If subchondral cysts are especially prominent, consider PVNS or pseudogout.

(2) If patellofemoral disease is prominent, with little or no medial or lateral compartment disease, consider pseudogout and look for chondrocalcinosis.

9. Ankle: OA rare in the absence of trauma.
10. Foot: Most common in first MTP and first tarsometatarsal (TMT).
11. Spine: Various manifestations are seen that have distinct terminology. These manifestations are often interdependent and therefore coexistent.

   a. *Degenerative disc disease:* Late manifestations are *decreased disc height and disc vacuum sign.* The disc may herniate posteriorly (seen by CT, magnetic resonance imaging (MRI), or myelogram), centrally into an adjacent vertebral body (*Schmorl's nodes*), or anteriorly, resulting in a *limbus vertebra* if it occurs prior to skeletal maturity. (A limbus vertebra results from separation of a ring apophysis from the underlying vertebral body by a herniated disc producing a separate triangular ossicle, usually located at the anterosuperior border of the vertebral body, seen on the lateral film.) Degenerative disc disease often results in reactive sclerosis of the adjacent end-plates termed *discogenic sclerosis* or *idiopathic segmental sclerosis.* Discogenic sclerosis may be difficult to differentiate radiographically from a disc space infection, but the end-plate remains intact in the former process. In addition, the sclerosis of the superior end-plate is distinctive, being triangle-shaped anteriorly. This should also be distinguished from blastic metastatic tumor.

   b. *Spondylosis deformans* results from bulging of the anulus fibrosus, stretching of Sharpey's fibers (the attachment of the anulus to the vertebral body), and *traction osteophyte formation* on the anterior and lateral vertebral body, arising several millimeters from the end-plates. These osteophytes may bridge the disc spaces in a bulky fashion, clearly distinct from the thin vertical syndesmophytes of AS. Since spondylosis deformans does not involve a joint, it is not a manifestation of OA but rather a degenerative disease of the spine.

c. *True osteoarthritis* involves the apophyseal joints (facets) and most commonly involves C_{5-7}, L_{4-5}, and S_1 vertebrae. Typical degenerative changes are seen (sclerosis and osteophyte formation) by plain film, but the extent and severity of the resultant stenosis is best judged by CT.[7] The facet OA often results in *spondylolisthesis without spondylolysis* and certainly *contributes to stenosis*.

d. *Uncovertebral* joints *(joints of Luschka)* are found only in the C_{3-7} *bodies*, located posterolaterally. Osteophytes may form and are seen on AP or lateral films, but neuroforaminal encroachment is best evaluated on oblique films.

e. *Spinal stenosis* may be seen on plain film as a decreased interpediculate distance or short AP pedicle length; however, these bony measurements have a very wide range of normal and, so, are useful only infrequently. Furthermore, soft tissue abnormalities often contribute to spinal stenosis (disc bulge, ligamentum flavum hypertrophy). The total picture of congenitally short pedicles modified by soft tissue abnormalities and facet OA is best evaluated by CT or MRI.

H. Bilateral symmetry may be present but is much less common than in RA.

I. Other features:
Erosive (inflammatory) OA: An arthropathy

found primarily in *middle-aged women* who experience *distinct inflammatory episodes,* similar to RA, with swollen red joints. ESR and rheumatoid factor are normal. *DIP and PIP* joints are involved, with loss of cartilage, sclerosis, and combined *erosive and productive bony changes*. The erosions on the proximal side of the joint tend to be central and the osteophytes, marginal, giving a distinctive *gull-wing* appearance.[8]

Fusion is seen occasionally.

The *differential diagnosis includes psoriatic arthritis, adult Still's disease, and septic joint (if monarticular). Typical OA of the first CMC joint and STT joints* (usually without erosions) *is often present* and is greatly helpful in arriving at the correct diagnosis.

J. *Differential diagnoses:* Discussed in Section G; varies according to the joint involved; includes RA, AS, psoriatic arthritis, PVNS, and pseudogout.

K. Suggested survey films: PA hands, AP pelvis, AP and lateral knees.

V. NEUROPATHIC (CHARCOT) JOINTS

> *Key Concepts:* Severe destructive arthropathy may be hypertrophic or atrophic. Distribution helps determine etiology: shoulder, syringomyelia; foot (talonavicular or TMT Lisfranc fracture dislocation), diabetes; knee, tabes dorsalis.

A. Definition: A severely destructive arthropathy, usually monostotic, with several etiologies. The *most common etiologies include diabetes, tabes dorsalis, and syringomyelia*. Other etiologies include myelomeningocele or spinal cord injury, multiple sclerosis, Charcot-Marie-Tooth disease, alcoholism, amyloidosis, intra-articular steroids,

congenital insensitivity or indifference to pain, and dysautonomia (Riley-Day syndrome). The primary *pathogenesis* is disputed, but most agree that an *initial alteration in sympathetic control of bone blood flow leads to hyperemia and active bone resorption. A neurotraumatic mechanism is usually important secondarily*, with a destructive cycle of (1) blunted pain sensation and proprioception; (2) relaxation of skeletal supporting structures and chronic instability; (3) recurrent injury with normal biomechanical stresses but abnormal joint loading; (4) bony fragmentation and joint disorganization. Progression may be rapid.
B. Epidemiology: Depends on etiology.
C. Clinical signs: Swollen, unstable joint. *The diagnosis may not be clinically obvious since 30% have pain* and the neurologic changes may be difficult to elicit.
D. Laboratory tests: Depends on etiology, VDRL, etc.
E. Extra-articular manifestations: Depends on etiology. There may be typical findings of syphilis or diabetes. A syrinx may widen the cervical canal. Congenital insensitivity or indifference to pain may be manifest by multiple scars on hands and by intraocular foreign bodies.
F. General radiographic description: The classical description is of *five Ds*—increased *density*, joint distension, bony debris, joint disorganization, and dislocation. This describes the *hypertrophic* variety well, but only 20% of Charcot joints are purely hypertrophic. Forty percent *are primarily atrophic* with such distinctive bone resorption that there often is a sharp transverse cut-off of the bone in the metadiaphysis, with complete resorption of the articular portion, that appears almost surgical. Another 40% are combined hypertrophic and atrophic.[9]
   1. Soft tissue: *Large effusions.*
   2. Abnormal calcifications: Much *bony debris and*

fragmentation. The effusions may decompress down fascial planes, carrying bony debris far from the joint itself.

3. *Bone density: Normal to increased* (except in diabetics).

4. *Cartilage destruction:* Prominent early finding.

5. *Erosive and productive changes coexist;* One or the other may predominate, depending on whether it is an atrophic or hypertrophic form.

6. Subchondral cysts: May be present.

7. Periostitis or enthesopathy: Not present.

8. Ankylosis: Rare; in fact, surgical arthrodesis is difficult.

9. Ligamentous abnormality: Prominent finding, with *laxity and joint subluxation or dislocation.*

G. Joints most commonly affected: In many cases, the etiology of the neuropathic joint may be suggested by which joint is involved.

1. *Knee* involvement is usually due to *tabes dorsalis,* which may also involve other joints of the lower extremity or the lumbar spine. Changes are usually hypertrophic.

2. *Foot* involvement is most commonly due to *diabetic arthropathy,* specifically if the *talonavicular* and TMT joints are involved (the latter often *resembles a Lisfranc fracture dislocation.* Diabetic neuroarthropathy in fact is a more common etiology for Lisfranc fracture dislocation than is trauma). These are often the atrophic or mixed form of Charcot joints. Superimposed infection is a diagnostic problem.

3. *Shoulder* neuroarthropathy is most commonly secondary to *syringomyelia.* It is almost always *atrophic,* with resorption of most or all of the humeral head and neck.

4. *Spine* neuroarthropathy may be seen in patients with *spinal cord trauma* who have

undergone instrumentation and fusion. The *first mobile segment* adjacent to the fusion is most commonly involved.

H. Bilateral symmetry: Rare.

I. Other features: In patients with decreased pain sensation an undetected fracture may occur (probably initiated as a stress fracture). Exuberant periostitis and callus formation may even suggest a bone-forming tumor.

J. Differential diagnoses:
   1. Hypertrophic form: Severe osteoarthritis.
   2. Atrophic form: Joint sepsis.

K. Survey films: Based on clinical suspicion.

VI. BIOCHEMICAL ABNORMALITIES
Gout

> **Key Concepts:** Long interval between onset of clinical disease and radiographic findings: Para-articular erosions with overhanging edges and adjacent tophi with amorphous calcifications are pathognomonic. Bone density is normal, and erosion margins, sclerotic. Hands and feet most commonly are involved.

A. Definition: A *sodium urate* crystal-induced synovial inflammation that may be limited to occasional acute attacks or may be a chronic arthropathy with crystal deposition in capsular and synovial tissues, periarticular soft tissues, articular cartilage, and subchondral bone. This crystal deposition and inflammatory reaction provoke very specific degenerative changes. Disease is (1) idiopathic, (2) secondary to enzyme defects, or (3) secondary to chronic disease processes such as myeloproliferative disorders, renal disease, hyper- or hypoparathyroidism, psoriasis, or diuretic therapy.

B. Epidemiology:
   1. Occurs mostly in *middle-aged or elderly males* (extremely rare in premenopausal women), with fairly strong hereditary factors.
   2. Onset at an early age is most often due to renal disease or a myeloproliferative disorder.
   3. Caucasians affected more commonly than blacks.
   4. *Gouty attacks typically occur for several years* (average 12) *before radiographic abnormalities are seen.*

C. Clinical signs: Red, swollen, *extremely painful joints;* usually monarticular or oligoarticular.

D. Pertinent laboratory tests:
   1. Hyperuricemia determination.
   2. *Sodium urate crystals* may be seen in synovial fluid by polarizing microscope.

E. Extra-articular manifestations:
   1. *Tophus in bursa* (olecranon or prepatellar bursa most commonly).
   2. Tophus in the helix of the ear and occasionally other soft tissue sites.

F. General radiographic description:
   1. *Soft tissue* alterations: *Tophi* are eccentric soft tissue nodules adjacent to a joint, often containing amorphous calcification.
   2. Abnormal calcifications: Amorphous calcification within tophi; occasional focal increased density within bone. Chondrocalcinosis may occur but is much less common than in pseudogout.
   3. *Bone density: Normal,* with osteoporosis only in extremely long-standing disease due to disuse.
   4. Cartilage destruction: *Cartilage often remains intact, even late in the disease process and with adjacent erosions.* This may be an important discriminating feature.
   5. Erosive versus productive changes: *Erosions* may be *intra-articular (often marginal) or para-*

articular (often beneath tophi). The para-articular erosions are almost pathognomonic for gout and may show the additional feature of an *"overhanging edge,"* a bony lip or excrescence extending out toward the tophus, beyond the normal bony margin. Occasionally, with long-standing disease, the erosions may be as mutilating as those of psoriatic arthritis or RA. One helpful feature is that the *erosions usually have a sclerotic margin. Productive changes* are in the form of enlargement of the ends of phalanges and secondary osteoarthritis with osteophyte formation.

6. Subchondral cysts: Not present, but the sclerotic marginated erosions give a similar appearance.
7. Enthesopathy: May be present.
8. Ankylosis: Very rare.
9. Ligamentous abnormality: Occasional ligament rupture.

G. Joints most commonly affected: In general, *lower extremity is more commonly involved than upper extremity,* and *small joints,* more often than large joints.

1. *Foot and ankle:* First MTP most common; other MTPs, IPs, midfoot, and hindfoot may also be involved.
2. *Knee:* May see marginal erosions or erosions on the articular surface of the *patella* (which may simulate neoplasm or the normal variant dorsal defect of the patella).
3. *Hand and wrist:* DIPs, PIPs, intercarpal joints are most commonly involved; MCPs less frequently.
4. Elbow: Extensor surface disease (olecranon).
5. SI joint involved occasionally.

H. Symmetry: Asymmetric.
I. Other features: *Tophi and erosions are rarely seen until after several years of chronic disease.*

J. Differential diagnoses:
   1. Early, nonclassic gout may be mistaken for rheumatoid or psoriatic arthritis.
   2. Gout may be confused clinically with pseudogout, but site of involvement should help differentiate the two.
   3. Xanthomatosis gives soft tissue masses on tendons, sometimes with associated bony erosion.
K. Suggested survey films: Feet: AP, lateral; hands: PA.

Calcium Pyrophosphate Dihydrate (CPPD) Crystal Deposition Disease

> *Key Concepts:* Intra-articular crystal deposition; pyrophosphate arthropathy often demonstrates chondrocalcinosis and a pattern of degenerative joint disease (DJD) in very specific joints (knee, wrist, second and third MCPs, hip, elbow) as well as in a specific distribution within a joint (e.g., patellofemoral compartment in knee, radiocarpal compartment in wrist). Subchondral cysts are a very prominent feature.

A. Definition:
   1. A relatively *common* arthropathy resulting from CPPD crystal deposition (usually intra-articular) with *radiographically distinctive features.*
   2. Various terminology relating to this disease is often loosely and incorrectly applied. Resnick suggests the following strict terminology:
      a. *Chondrocalcinosis: Nonspecific cartilage calcification;* the crystals deposited usually are, but may not be, CPPD, and there may or may not be an associated arthropathy.
      b. *CPPD crystal deposition disease:* More

specific form of chondrocalcinosis due to CPPD crystal deposition; there may or may not be an associated arthropathy.

c. *Pyrophosphate arthropathy: Pattern of structural joint damage occurring in CPPD deposition disease that has the appearance of DJD.* However, it occurs in *specific sites that are unusual for DJD but specific for pyrophosphate arthropathy.* Although pyrophosphate arthropathy is associated with CPPD deposition, it *may or may not be manifest as chondrocalcinosis. Pyrophosphate arthropathy may or may not present clinically as pseudogout.*

d. *Pseudogout:* The *clinical presentation of CPPD deposition disease which resembles gout* (intermittent acute attacks). This is only one of several clinical manifestations of CPPD deposition disease but is often used loosely in lieu of "pyrophosphate arthropathy" in describing the radiographic findings of the latter disease.

B. Epidemiology:
1. *Middle aged or elderly male or female.*
2. May coexist with gout.

C. Clinical signs: Several common patterns.
1. Pseudogout (10% to 20%): Acute self-limited attacks simulating gout or infection.
2. Pseudo-RA (2% to 6%): More continuous acute attacks simulating RA.
3. Pseudo-DJD (35% to 60%): Chronic, progressive arthropathy but with acute exacerbations.
4. Pseudo-DJD without acute exacerbations (10% to 30%).
5. Asymptomatic (10% to 20%).
6. Pseudoneuropathic (rare): Rapidly destructive form.

D. Laboratory tests:
1. Joint aspiration, with calcium pyrophosphate

crystals seen in synovial fluid under polarizing microscope.
2. No associated biochemical abnormality.
E. Extra-articular manifestations: None.
F. General radiographic description of pyrophosphate arthropathy:
  1. Soft tissue alterations: Local swelling.
  2. Abnormal calcifications: Usually intra-articular.
     a. *Chondrocalcinosis is seen in either hyaline or fibrocartilage, most commonly in the knee (menisci), triangular fibrocartilage complex of the wrist, symphysis pubis, and acetabular labrum.* Other sites include the glenoid labrum, acromioclavicular joint, sternoclavicular joint, and anulus fibrosus.
     b. Crystals may also be deposited in the synovium, capsule, tendons, and ligaments.
  3. *Bone density: Normal.*
  4. Cartilage destruction: Present.
  5. Erosive versus productive bony change: *primarily productive*, with *sclerosis*, osteochondral fragments, and osteophytes (especially notable at MCP heads).
  6. *Subchondral cysts:* A distinctive feature, they are *common and tend to be very large, sometimes simulating neoplasm.*
  7. Periostitis or enthesopathy: Not present.
  8. Ankylosis: Rare.
  9. Ligamentous abnormalities: Occasional crystal deposition; instability pattern may be seen in the wrist (see G, below).
G. *Joints most commonly affected: The most distinctive feature of pyrophosphate arthropathy—knee, wrist, and second and third MCPs are involved most often.*
  1. *Knee: Chondrocalcinosis in menisci as well as hyaline cartilage.* The degenerative features may be seen in all three compartments but are *often much more prominent in the*

patellofemoral compartment than the medial or lateral compartment.

2. *Wrist (Fig 2–10): Chondrocalcinosis in the TFCC and/or intraosseous ligaments. Degenerative changes are specifically found in the radiocarpal joint,* often with scapholunate dissociation and scaphoid erosion into the distal radial articular surface. Proximal migration of the capitate between the dissociated scaphoid and lunate may result in a scapholunate advanced collapse (SLAC) wrist pattern. This distribution is significantly different from that seen in typical DJD.

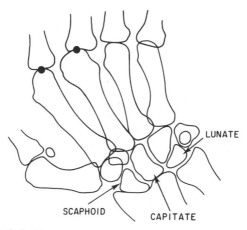

FIG 2–10.
Calcium pyrophosphate arthropathy pattern in the hand: radiocarpal disease, with scapholunate dissociation and SLAC wrist deformity; arthropathy with prominent osteophytes is seen typically at the second and third MCP joints.

3. Hand: *Second and third MCPs* specifically involved; IPs are spared.
4. Hip: If large subchondral cysts are seen with "DJD," consider pyrophosphate arthropathy and look for labral chondrocalcinosis.
5. Shoulder and elbow: A DJD pattern should suggest pyrophosphate arthropathy since DJD is rare in these joints in the absence of trauma. Chondrocalcinosis should be sought.

H. Symmetry: Often bilateral but not necessarily symmetric disease.

I. Other features: Several other diseases are seen in association with CPPD crystal deposition: DJD, trauma, gout, hyperparathyroidism, hemochromatosis, Wilson's disease (documentation questionable), ochronosis.

J. Differential diagnoses:
   1. For chondrocalcinosis:
      a. Other diseases associated with CPPD deposition (see I, above).
      b. Scleroderma.
      c. HA crystal deposition.
      d. Periarticular calcification: Renal osteodystrophy, collagen vascular disease, vitamin D metabolic abnormalities.
   2. For the arthropathy:
      a. DJD, but site of involvement differs significantly and reliably.
      b. Others, such as RA or gout, may be suggested clinically but not radiographically since pyrophosphate arthropathy is not an erosive disease.

K. Survey films:
   1. For chondrocalcinosis: AP knees, PA hands and wrists, AP symphysis pubis.
   2. For arthropathy: Add lateral knees to films for chondrocalcinosis, above.

Hemochromatosis Arthropathy

> **Key Concepts:** Nearly identical arthropathy to that of pyrophosphate arthropathy, with degenerative-type changes at specific sites: patellofemoral predominance at knee; radiocarpal and second or third MCP joints in hand. Osteoporosis may be present.

A. Definition: An arthropathy that develops in up to *50% of patients who have hemochromatosis,* presumably due to *accumulation of iron and/or CPPD crystals in joints.* Hemochromatosis itself may be either primary (increased GI absorption of iron) or secondary (increased intake of iron by means of blood transfusions, alcoholism, or excess ingestion).

B. Epidemiology:
  1. Age: *Onset in middle age.*
  2. Sex: *Males* affected much more often than females.

C. Clinical signs: Usually mild pain, swelling, and stiffness, but acute attacks may occur.

D. Laboratory tests: Increased serum iron and iron-binding capacity.

E. Extra-articular manifestations: *Clinical triad of bronze skin, cirrhosis, and diabetes.* Heart disease is also seen.

F. General radiographic description of arthropathy: *Nearly identical to pyrophosphate arthropathy.*
  1. Soft tissues: Mild swelling.
  2. Abnormal calcifications: *Chondrocalcinosis* is common but not always associated with joint disease.
  3. Bone density: *Osteoporosis* common.
  4. Cartilage destruction: Present focally.

  5. Erosive versus productive changes: Primarily *productive, with large beaklike osteophytes,* noted especially on MCP heads.
  6. *Subchondral cysts: Prominent,* as in pseudogout arthropathy.
  7. Periostitis or enthesopathy: None.
  8. Ankylosis: None.
  9. Ligamentous abnormality: No deformity.
G. *Joints most commonly affected: Same as pyrophosphate arthropathy.*
  1. *Second and third MCPs.*
  2. *Radiocarpal joint.*
  3. *Knee, especially patellofemoral compartment.*
  4. Hip.
  5. Shoulder.
H. Symmetry: Usually present.
I. Other features: None.
J. Differential diagnosis: Pyrophosphate arthropathy: distinctive features of hemochromatosis include osteoporosis, especially prominent beaklike osteophytes on MCP heads and involvement of more carpal areas.
K. Survey films: PA hands and wrists; AP and lateral knees.

Wilson's Disease

> *Key Concepts:*   Osteopenia in young males is suggestive; irregular subchondral bone and fragmentation resembling osteochondritis.

A. Definition: An autosomal-recessive disease associated with abnormal accumulation of *copper.*
B. Epidemiology:
  1. *Males* more often affected than females.
  2. Manifest in adolescents and young adults but arthropathy may develop later.
C. Clinical signs: The articular abnormalities are often asymptomatic. Otherwise, mechanical degenerative symptoms are present.

D. Laboratory tests: Evidence of renal tubular abnormalities.
E. Extra-articular manifestations: Basal ganglia degenerative changes, cirrhosis, Kayser-Fleischer rings in cornea.
F. General radiographic description of the arthropathy:
   1. Soft tissues: Normal.
   2. Abnormal calcification: *Chondrocalcinosis* may be present. There may be an association of Wilson's disease and CPPD crystal deposition disease due to pyrophosphatase inhibition by copper.
   3. Bone density: *Osteopenic,* perhaps due to renal tubular disease.
   4. Cartilage destruction: Present, indicated by decreased width.
   5. Erosive versus productive change: *Indistinct, irregular subchondral bone, with several small fragments or ossicles; this may give the appearance of osteochondritis.* Sclerosis may be present.
   6. Subchondral cysts: Present.
   7. Periostitis or enthesopathy: May be present.
   8. Ankylosis: None.
   9. Ligamentous abnormality: None.
G. Joints most commonly affected: *Wrist and hand (especially MCPs),* foot, hip, shoulder, elbow, knee.
H. Bilateral symmetry: Often present.
 I. Other features: None.
 J. Differential diagnoses:
   1. CPPD/hemochromatosis: *The distribution of the arthropathy is the same but the irregularity and fragmentation of the subchondral bone is distinctive.* Chondrocalcinosis may be present in both.
   2. *DJD:* Osteopenia as well as joint distributions make DJD unlikely.
K. Survey films: PA film of hands.

Calcium Hydroxyapatite (HA) Deposition Disease

> *Key Concepts: Periarticular calcification; shoulder most common site; underlying bone usually normal.*

A. Definition: Periarticular calcifications due to HA crystal deposition, generally monarticular, and with an inflammatory reaction but usually without structural joint abnormality. Etiology is unknown, but may relate to repeated minor trauma and deposition in necrotic tissue.

B. Epidemiology: Middle to old age; affects men and women equally.

C. Clinical signs: Generally painful single joint.

D. Laboratory tests: HA crystals seen by light microscopy with Wright's stain or by electron microscopy, not via polarizing microscope.

E. Extra-articular manifestations: None.

F. General radiographic description:
  1. Soft tissue: May be swollen.
  2. Abnormal calcification: The finding that suggests the diagnosis: The *calcification* is usually *cloudlike and homogeneous,* occurring *periarticularly (tendons, ligaments, capsule, bursa).* Occasionally it is intra-articular, producing chondrocalcinosis.
  3. Bone density: Normal to partly sclerotic.
  4. Cartilage destruction: Usually none.
  5. Erosive or productive bone changes: None.
  6. Subchondral cysts: Small cysts may be seen.
  7. Periostitis or enthesopathy: None.
  8. Ankylosis: None.
  9. Ligamentous abnormality: Calcification.

G. Joints most commonly affected:
1. *Shoulder: Calcification located at sites of tendon insertion, i.e., along the greater tuberosity (supraspinatus, infraspinatus, teres minor), lesser tuberosity (subscapularis), origins of biceps tendons, as well as subacromial and subdeltoid bursae.* The complex is often called the *Milwaukee shoulder.*
2. Elbow: Triceps and collateral ligament insertions.
3. *Wrist:* Any tendon insertion, but especially that of the *flexor carpi ulnaris* adjacent to the pisiform.
4. *Hand: Periarticular deposits around the MCPs and IPs.*
H. Bilateral symmetry: Not expected.
I. Other features: None.
J. Differential diagnoses: Other conditions in which HA crystal deposition occurs:
1. Renal osteodystrophy.
2. Hypoparathyroidism.
3. Hypervitaminosis D.
4. Milk-alkali syndrome.
5. Collagen vascular disease.
6. Tumoral calcinosis.
7. Dystrophic calcification secondary to inflammation or tumor.
8. Myositis ossificans (progressiva or secondary to trauma).
9. Parasitic calcification.
10. Enthesopathy.
11. Gouty tophus.
12. CPPD crystal deposition.
K. Survey film: AP film of shoulder and other symptomatic sites.

Alkaptonuria (Ochronosis)

Key Concepts: Dystrophic dense calcification, most commonly involving the vertebral discs, with degenerative disc signs. Extraspinal manifestations resemble those of DJD.

A. Definition: A hereditary metabolic abnormality due to absence of homogentisic acid oxidase and consequent accumulation of homogentisic acid in various organs, including connective tissues.
B. Epidemiology:
   1. Sex: Affects males and females equally.
   2. Age: Pigmentation occurs in early adulthood and arthropathy, later.
C. Clinical signs: Mild pain and decreased range of motion, usually not requiring intervention.
D. Laboratory tests: Urinary homogentisic acid.
E. Extra-articular manifestations: Pigmentation seen in helix of ear.
F. General radiographic description:
   1. Soft tissues: Normal.
   2. Abnormal calcifications: *Dystrophic (HA crystal) calcification most commonly involving the discs, but also cartilage, tendons, and ligaments.*
   3. Bone density: *Osteoporotic.*
   4. Cartilage destruction: Becomes brittle and fragmented.
   5. Erosive versus productive changes: *Sclerosis and small osteophytes.*
   6. Subchondral cysts: Present.
   7. Periostitis and enthesopathy: Not present.
   8. Ankylosis: None.
   9. Ligamentous abnormality: Calcification and rupture.
G. Joints most commonly affected:
   1. *Spine: Osteoporotic with dense disc calcification.*
   2. Extraspinal: *SI joints, symphysis pubis, and large peripheral joints may be involved (usually after spine) and show changes of mild DJD.*

H. Bilateral symmetry: Often present.
I. Other features: None.
J. *Differential diagnosis:*
   1. *Dystrophic calcification of the nucleus pulposus:* AS or CPPD.
   2. DJD: Osteophytes are not as prominent in ochronosis.
K. Survey films: Lateral spine.

VII. MISCELLANEOUS DISORDERS
Pigmented Villonodular Synovitis (PVNS)

> *Key Concepts:* Monarticular, hemorrhagic effusions, often with erosions, most often occurring in knee or hip.

A. Definition: A proliferative disorder of the synovium of unknown etiology; it may be extra-articular (giant cell tumor of the tendon sheath, seen as soft tissue swelling with or without adjacent bony erosions) or intra-articular, which is described in this section.
B. Epidemiology: Middle to old age.
C. Clinical signs: Painless effusion.
D. Laboratory tests: None.
E. Extra-articular manifestations: *Giant cell tumor of the tendon sheath is histologically indistinguishable from intra-articular PVNS.*
F. General radiographic description:
   1. Soft tissues: *Hemorrhagic effusion.*
   2. Abnormal calcifications: Very rare calcific metaplasia.
   3. Bone density: Normal to mild osteoporosis.
   4. *Cartilage: Preserved until late in the disease process.*
   5. Erosive versus productive changes: *Erosions on both sides of the joint are common findings.*
   6. Subchondral cysts: May be present.
   7. Periostitis or enthesopathy: Absent.

    8. Ankylosis: Absent.
    9. Ligamentous abnormality: Absent.
G. *Joints most commonly affected: Knee, hip, elbow.*
H. Bilateral symmetry: *Usually monarticular.*
 I. Other features: Filling defects on arthrography.
 J. Differential diagnoses: Infection and other
    monarticular arthritides.
K. Survey films: Only as clinically suggested.

Synovial Chondromatosis

> **Key Concepts:** Synovial metaplasia producing
> multiple round cartilaginous or osseous intra-arti-
> cular loose bodies, most often in the knee or hip.

A. Definition: A *synovial metaplasia* of unknown
    etiology in which *cartilaginous nodules* arise from
    projections of the synovium. If the pedicle
    remains intact and a blood supply continues,
    osseous bodies may form. If the bodies become
    loose, cartilaginous ones which are nourished by
    the synovium may continue to grow, whereas
    osseous bodies may resorb and grow only if they
    become reattached to the synovium.
B. Epidemiology: Males affected more often than
    females; third to fifth decades.
C. Clinical signs: Usually pain and limited range of
    motion, though it may be asymptomatic.
D. Laboratory tests: None.
E. Extra-articular manifestations: None.
F. General radiographic description:
    1. Soft tissues: *Usually no effusion.*
    2. Abnormal calcifications: *Generally multiple*
       round bodies similar in size and variable in
       mineralization, sometimes appearing lamellated
       and even trabeculated.
    3. Bone density: Normal.
    4. *Cartilage: Normal, though eventually loose bodies*
       may lead to secondary mechanical cartilage
       destruction.

5. Erosive versus productive change: Loose bodies may cause mechanical erosions and may also eventually promote DJD with concomitant productive changes.
6. Subchondral cysts: Only with secondary DJD.
7. Periostitis or enthesopathy: None.
8. Ankylosis: None.
9. Ligamentous abnormality: None.

G. *Joints most commonly affected: Knee, hip, elbow, shoulder.*
H. Bilateral symmetry: *Monarticular.*
I. Other features: Filling defects on arthrography.
J. Differential diagnoses: Loose bodies in DJD.
K. Survey films: As indicated clinically.

Osteochondroses

> **Key Concepts:** Fragmentation of epiphyseal or apophyseal surfaces, which are usually convex; most commonly affects skeletally immature patients. Trauma is the most common etiology. Differential diagnosis usually includes normal variants.

A. Definition: *The osteochondroses represent an inelegant and artificial grouping of disease processes, all once believed to represent necrosis. Only a few are true necroses; most are traumatically induced. Several are normal variants included only because they are "named" disease processes.*
B. Epidemiology:
   1. Age: Skeletally immature persons or young adults.
   2. Sex: Males more often affected than females (except Freiberg's necrosis, see below).
C. Clinical signs: Pain, decreased range of motion.
D. Laboratory tests: None.
E. Extra-articular manifestations: None.
F. *General radiographic description: Increased density, with bony fragmentation in a lucent bed; flattening of*

a convex surface may occur, though the overlying cartilage is usually intact. Secondary degenerative change may occur much later.

G. Joints affected:
  1. *Hip: Legg-Calvé-Perthes* disease:
    a. *A true necrosis, seen at ages 4 to 8, when the vascular supply to the femoral head is most at risk.*
    b. *Males affected more often than females.*
    c. *Rare in blacks.*
    d. Asymmetric, *10% bilateral.*
    e. *First radiographic sign* may be *effusion.* Later, *fragmentation and flattening* of the ossification center develop. Metaphyseal irregularity and "cysts" are manifestations of a growth abnormality that results in a short, wide femoral neck.
    f. *Older patients* at time of diagnosis and *females* (skeletally more mature) have a *poorer prognosis.* Gage advocated several "head at risk" signs, but all relate to *lateral extrusion of the femoral head ossification center* (calcification lateral to the acetabular rim), which indicates a lack of coverage of the femoral head and *has a poor prognosis since the head and acetabulum will not be congruent.* Surgical treatment is varied, but the objective is to contain the femoral head within the acetabulum, thus attaining congruity as the hip matures.
    g. *Differential diagnosis is hypothyroidism* since the "cretinoid hip" gives a similar appearance.
  2. *Lunate malacia: Keinbock's* disease: A true necrosis, believed to be related most often to trauma but commonly also to ulnar minus variant.
  3. *Freiberg's* disease:
    a. A true necrosis, involving the *second and third metatarsal heads* most commonly

(fourth and first less commonly).

   b. More common in teenage females; may be related to the trauma of bearing weight in high-heeled shoes.

4. *Knee: Osteochondritis dissecans affects the lateral portion of the medial femoral condyle almost exclusively.*

   a. Believed to be related to trauma.

   b. Must be differentiated from *spontaneous necrosis*, which affects an *older* patient, has an *acute onset* with *severe pain,* demonstrates *flattening on the weight-bearing (more medial) aspect of the medial femoral condyle,* and may be associated with a medial meniscal tear.

5. *Tibial apophysis: Osgood-Schlatter's disease.*

   a. Several ossification centers may normally be seen in this apophysis, so *soft tissue swelling and pain should be present in addition to bony fragmentation* to suggest this disease process.

   b. Thought to be a traumatically induced avulsion.

6. Sinding-Larsen-Johanssen disease: Avulsion and fragmentation of the inferior pole of the patella in this same age group.

7. *Blount's* disease: Osteochondrosis of the *medial aspect of the proximal tibial epiphysis, with resultant genu varus.*

   a. *Infantile type* is *most common,* especially among *blacks.* It *evolves when the normal physiologic bowing of the lower extremity worsens with weight bearing,* especially in early walkers. There is no true necrosis, but the persistent microtrauma of weight bearing and abnormal pressure causes fragmentation of the medial metaphysis, which in turn causes epiphyseal deformity. More than 50% bilaterality.

   b. Adolescent type is unilateral and relates to

trauma or infection causing bony bridging of the medial growth plate.

8. *Talar dome osteochondritis dissecans:* Develops on medial or lateral convex surfaces of the talus and is probably related to trauma and *ankle laxity.*

9. *Kohler's disease: A dense, fragmented tarsal navicular may be a normal variant* of ossification that is a self-limited process and proceeds to normal ossification; Thus, Kohler's disease should be suggested only when there is fragmentation of a previously normal tarsal navicular in the presence of pain.

10. *Panner's disease: Capitellum,* most commonly seen in adolescent fast-ball pitchers; self-limited with normal regeneration.

11. *Scheuermann's* disease: Strict definition requires abnormalities in *three contiguous vertebral bodies with at least 5 degrees anterior wedging* in each, resulting in a *dorsal kyphosis.* (Normal kyphosis is 20 to 40 degrees, measured between T_4 and T_{12}). *The end-plates are usually irregular* and often associated with Shmorl's nodes. The etiology is unknown but is not necrosis. Hypotheses include congenital end-plate weakness. *Males and females* are affected equally, especially teenagers; *Lower thoracic spine* is involved in 75%.

12. *Normal variants* that have fragmented ossification centers and *simulate osteochondroses:*
    a. *Femoral condyle* (more posteriorly located than osteochondritis dissecans, seen best on notch view) in skeletally immature persons.
    b. *Calcaneal apophysis* ("Sever's disease").
    c. *Trochlea.*
    d. *Lateral epicondyle.*

e. *Anterior tibial apophysis* (Osgood-Schlatter's disease).
f. Tarsal navicular (Kohler's disease).
g. Ischiopubic synchondrosis (Van Neck's disease).

H. Bilateral symmetry: Rare.
I. Other features: Arthrography usually demonstrates intact cartilage and, therefore, development of loose bodies is rare.
J. Differential diagnoses:
   1. Normal variants (see above).
   2. Vascular events.
   3. Epiphyseal dysplasias.

Hypertrophic Osteoarthropathy (HOA)

> *Key Concepts:* Presents clinically as arthritis; radiographically has normal joints but thick periosteal reaction; may be primary or secondary.

A. Definition: A disease process of unknown etiology that *presents clinically as arthritis* with painful, swollen joints. Radiographically, the *joints appear nearly normal:* there are only soft tissue swelling and occasional effusions, no erosive or productive changes. The major abnormality is found in the corners of the film— a *symmetric periosteal reaction.* The reaction may show onion skinning, irregularity, or waviness; its *thickness and extent probably depend on the duration of the disease.*

B. HOA may be primary or secondary:
   1. Primary HOA (pachydermoperiostitis) is a spectrum of diseases ranging from mere periostitis to the complete process of periostitis, clubbed digits, and thickening of the skin (facial and hands), resulting in pawlike hands. It is often familial and is much more common in males than females.

The onset is in adolescence, and there is usually spontaneous arrest of the process in young adulthood.

2. Secondary HOA is periostitis in the extremities noted in patients with a number of disease processes (usually intrathoracic): bronchogenic carcinoma (most common), other malignant, benign, or chronic suppurative disease of the lung, cyanotic heart disease, liver or biliary cirrhosis, and IBDs. The mechanism of the reaction is entirely unknown. Interestingly, a thoracotomy may lead to clinical remission almost immediately, with slower radiographic resolution.

C. Differential diagnosis of the periostitis:

1. Although it has been observed that primary HOA may have irregular excrescences extending to the epiphyses, it is unlikely that primary and secondary HOA can be differentiated radiographically in the individual case.

2. *Thyroid acropachy:* Generally a feathery, spiculated reaction, predominantly on the hands and feet; clubbing of the digits is often present.

3. *Vascular insufficiency* may also produce symmetric periosteal reaction, usually restricted to the lower limbs.

Avascular Necrosis (AVN)

Key Concepts: Early changes of sclerosis, followed by subchondral fracture, bony fragmentation, flattening of weight-bearing surfaces, but cartilage remains intact.

A. Definition: Necrosis of bone, restricted, in this section, to epiphyseal locations. Etiologies are usually related to trauma (vessel interruption), vascular compression, or intraluminal obstruction. The following etiologies are listed in approximate descending order of occurrence:

1. *Idiopathic.*
2. *Trauma:* Due to delayed reduction of a *dislocation or to subcapital fracture.*
3. *Steroids:* Exogenous or endogenous (Cushing's disease); thought to be due to an increase in fat cell size and resultant increased pressure.
4. *Alcoholism:* May be due to fat emboli and increased marrow pressure.
5. *Sickle cell disease:* Vascular occlusion from sickled cells.
6. *Gaucher's* disease: Sinusoids are packed with Gaucher's cells.
7. *Caisson* disease: Nitrogen embolization following rapid decompression, causing an increase in marrow pressure.
8. *Radiation:* Direct toxic effect on vascular supply to bone.
9. SLE: The vasculitis may be additive to the steroid therapy in causing AVN. AVN in unusual sites (talus, humerus) should suggest SLE as an etiology.
10. Pancreatitis: Fat necrosis as well as probable relationship to alcoholism.

B. Epidemiology: Relates to etiology.
C. Clinical signs: Pain with weight bearing, decreased range of motion.
D. Laboratory tests: None.
E. Extra-articular manifestations: None.
F. General radiographic description:
   1. *Cartilage is normal* since it is nourished by synovial fluid.
   2. *First radiographic sign is sclerosis,* which is due to different processes at different times. *Initially* it is a *relative sclerosis* (necrotic bone initiates an inflammatory response in the surrounding vascularized bone; the resultant

hyperemia causes osteoporosis). Later, a
reactive interface develops, with bone
formation causing increased density.
3. Later may develop a linear subchondral
fracture (*crescent sign*), best seen on a frog-leg
film. With further subchondral fragmentation,
flattening and *bone deformity* occur. Sclerosis
may be increased due to impacted fragments.
4. Still later, with revascularization, repair and
remodeling in the avascular segment occurs;
this "*creeping substitution*" also causes
sclerosis.
5. Secondary DJD may occur later.
G. Sites of most common occurrence: *Femoral head,
lunate, proximal pole of the scaphoid, and body of the
talus* are all at risk for posttraumatic AVN.
H. Bilateral symmetry: With systemic processes,
often bilateral but usually not symmetric.
I. Other features:
1. MRI is the most sensitive imaging tool for
detection of AVN[10, 11] but is nonspecific.
2. Radionuclide bone scan initially shows a
photon deficit, followed by increased activity
with revascularization and repair.
J. Differential diagnosis: DJD may simulate
advanced AVN of the hip since a large
inferomedial osteophyte may make the head
appear flattened; intact cartilage should
differentiate AVN from DJD until the AVN
develops secondary DJD.
K. Survey films for AVN from systemic causes: AP
and frog-leg pelvis films (the frog-leg view is
usually more sensitive).

Dish: Diffuse Idiopathic Skeletal Hyperostosis (Forestier's Disease)

> **Key Concepts:** Severe productive changes of the soft tissues surrounding the vertebral bodies, forming large bulky, flowing osteophytes; predominantly in the thoracic spine, though the entire column may be involved. Clinical signs are mild compared to the radiographic appearance. Strict radiographic criteria should be adhered to in order to avoid an incorrect diagnosis.

A. Definition: *Severe productive bone changes of the spine,* including the *anulus fibrosus, anterior longitudinal ligament,* and even *paravertebral connective tissues,* of unknown etiology. *Three strict radiographic criteria* must be fulfilled to suggest the diagnosis:[12]
   1. Flowing ossification of the anterolateral aspect *involves at least four contiguous vertebral bodies.* (Note that this rules out spondylosis deformans.)
   2. There is relative *preservation of disc height* over the involved segments, and degenerative disc disease (as manifest by vacuum signs or discogenic sclerosis) is not present.
   3. *Sacroiliitis and facet ankylosis are not present* (eliminating ankylosing spondylitis from consideration).
B. Epidemiology: *Very common abnormality* in middle-aged and elderly persons; *males* more commonly affected than females.

C. Clinical signs: *Milder than the extent of ossification suggests:* mild back pain, little deformity, and mild decrease in range of motion.
   a. With cervical disease, may develop dysphagia.
   b. More peripherally, may have tendinitis.
D. No laboratory abnormality.
E. Extra-articular manifestations: None.
F. General radiographic description:
   1. Soft tissue alterations: Calcification of the anterior longitudinal ligament, paravertebral soft tissue, and occasionally the posterior longitudinal ligament.
   2. Abnormal calcifications: See 1, above.
   3. Bone density: Normal.
   4. Cartilage destruction: None.
   5. No erosive changes: Purely productive; the osteophytes on the spine may be extremely large (up to 2 cm), irregular, and are located at the level of the disc. There may be lucent defects between the vertebral body and the ossifications at the discs, due to disc extrusion.
   6. Subchondral cysts: None.
   7. *Enthesopathy: Common,* especially in the *pelvis, calcaneus* (Achilles and plantar aponeurosis insertions), and anterior surface of the *patella* (quadriceps insertion).
   8. Ankylosis: Rare.
   9. Ligamentous abnormality: *Pelvic ligaments* may become *calcified* as may the anterior and posterior longitudinal ligaments.
G. Most common distribution:
   1. *Thoracic spine: Most commonly* involved; the osteophytes are usually found *anterolaterally, on the right side* (thought to be due to the pulsating aorta on the left).

2. *Cervical spine* is also commonly involved, *usually on the anterior aspect. Ossification of the posterior longitudinal ligament (OPLL) may be seen* and may cause neurologic symptoms. Such posterior ligament ossification may be concurrent with DISH or may be a solitary finding; there is an overlap between the two disease processes.

3. *Lumbar spine* may also be involved, especially $L_{1-3.}$

4. Superior (nonarticular portions) SI joints are often bridged by ligamentous calcification, but the lower two thirds of the SI joints is rarely affected.

H. Bilateral symmetry: Often absent in the thoracic spine.

I. Other features: None.

J. Differential diagnoses: Abiding by the strict radiographic criteria should eliminate the differentials:
   1. Spondylosis deformans.
   2. Discogenic sclerosis or degenerative disc disease.
   3. Ankylosing spondylitis.

K. Survey films: AP and lateral spine.

Transient Regional Osteoporosis

Definition: A disorder of unknown etiology that presents clinically as an arthritis and radiographically demonstrates only osteoporosis involving both sides of a joint. Cartilage remains normal, and there are neither erosive nor productive changes. Self-limited disease. Males affected more commonly than females; usually large joints of lower limbs; may be migratory. See also Chapter 4.

188 *Chapter 2*

REFERENCES

1. Schaller J: Spondyloarthritis and other forms of chronic arthritis affecting children, in Jacobs JC (ed): *New Frontiers in Pediatric Rheumatology.* New York, World Health Communications, Inc., 1986.
2. Bywaters E: Still's disease in the adult. *Ann Rheum Dis* 1971; 30:121–133.
3. Bjorkgren A, Pathria M, Sartoris D, et al.: Carpal alterations in adult-onset Still's disease, juvenile chronic arthritis, and adult-onset rheumatoid arthritis: Comparative study. *Radiology* 1987; 165:545–548.
4. Bjorkengren A, Resnick D, Sartoris D: Enteropathic arthropathies. *Radiol Clin North Am* 1987; 25:189–198.
5. Weissman BW, Rappaport AS, Sosman JL, et al: Radiographic findings in the hands in patients with systemic lupus erythematosus. *Radiology* 1978; 126:313.
6. Basset L, Blocka KL, Furst DE, et al: Skeletal findings in progressive systemic sclerosis (scleroderma). *AJR* 1981; 136:1121.
7. Pathria M, Sartoris D, Resnick D: Osteoarthritis of the facet joints: Accuracy of oblique radiographic assessment. *Radiology* 1987; 164:227–230.
8. Martel W, Stuck K, Dworin A, Hylland R: Erosive osteoarthritis and psoriatic arthritis: A radiologic comparison in the hand, wrist and foot. *AJR* 1980; 134:125–135.
9. Brower A, Allman R: Pathogenesis of the neurotrophic joint: Neurotraumatic vs neurovascular. *Radiology* 1981; 139:349.
10. Mitchell D, Kressel H, Argen P, et al.: Avascular necrosis of the femoral head: Morphologic assessment by MR imaging, with CT correlation. *Radiology* 1986; 161:739–742.

11. Mitchell D, Rao V, Dalinka M, et al.: Femoral head avascular necrosis: Correlation of MR imaging, radiographic staging, radionuclide imaging, and clinical findings. *Radiology* 1987; 169:709–715.
12. Resnick D, Niwayama G: Radiographic and pathologic features of spinal involvement in diffuse idiopathic skeletal hyperostosis (DISH). *Radiology* 1976; 119:559.

3

Trauma

GENERALIZATIONS

The radiologist confronts two dilemmas in the trauma patient. The first is recognition of injury. While this sounds simple, in practice, we quickly learn that it may indeed be quite difficult. It requires a detailed knowledge of gross anatomy and of normal variants that may be mistaken for fractures. The radiologist must become familiar with certain predictable fractures that occur in different age groups and with particular mechanisms that produce certain injuries. The radiologist armed with this information is unlikely to miss evidence of significant trauma.

In each of the subsequent trauma sections, categorized by anatomic site, the reader will find discussions of normal anatomy and of common trauma patterns. Because these are, in general, well-known facts and because this handbook is meant to be pedagogical, these discussions are not referenced exhaustively. If the reader wishes further detail, he is referred to Lee Rogers' excellent text, *Radiology of Skeletal Trauma.*

After the trauma is recognized, the radiologist must communicate his findings, using standard terminology, which is described below.

I. FRACTURE TERMINOLOGY

A. Definition and biomechanical principles.
   1. A fracture is a complete or incomplete break in the continuity of bone or cartilage.
   2. Bone is anisotropic; it has different mechanical properties when loaded in different directions. Adult cortical bone withstands the greatest stress in compression, less in tension, and the least with shear loading.
   3. When force is applied to bone, contraction of muscles alters the stress distribution and may allow the bone to support higher loads than expected.
   4. With greater speed of loading, more energy is stored before failure; with fracture, this energy is dissipated rapidly, resulting in extensive soft tissue damage.
   5. Repeated loading reduces the amount of total weight a bone can withstand owing to fatigue of muscles that normally redistribute the stress.
   6. A surgical defect (e.g., screw hole or site of bone resection) concentrates stress and decreases bone strength significantly: Such lesions are termed stress raisers.

B. Accurate terminology to be applied in all descriptions.
   1. *Open* vs. *closed* fracture.
   2. Incomplete vs. complete:
      a. *Incomplete* fractures most often are seen in children. The three major types are *torus* (a buckle of the cortex, a failure on the compressive side); *greenstick* (incomplete fracture on the tension side); and *plastic* (bending without angular deformity and without subsequent remodeling).
      b. *Complete* fractures generally can be described as *transverse*, *oblique*, or *spiral*.

3. *Comminution:* A fracture that produces more than two fragments. Subsets include *segmental* (two fracture lines isolating a discrete segment) and *butterfly* (a wedge-shaped separate fragment formed at the apex of the force). The latter terms should be used appropriately since they have implications for prognosis as well as treatment (e.g., the blood supply may be disrupted substantially, and a large butterfly fragment may lead to telescoping and instability of the fragments).
4. *Position.*
   a. Description of site, either by anatomic landmarks or by dividing long bones into thirds.
   b. A fracture located near an articular cortex should be designated either as intra- or extra-articular.
5. *Apposition:* Contact of the ends of the fracture fragments.
   a. *Anatomic.*
   b. *Displaced:* Anterior, posterior, lateral, or medial, or a combination of these; the degree of displacement is either measured or expressed as a percentage of cross-sectional diameter.
   c. *Lack of apposition:* Complete loss of contact of the bone ends. If the fragments overlap one another, leaving the shafts but not the ends in contact, it is termed *bayonet* apposition. Complete lack of apposition is termed *distraction* (most commonly due to excessive traction, interposed tissue, or resorption of fragment ends subacutely).
6. *Alignment:* Relationship of the long axes of the fracture fragments. *Angulation* is the loss of alignment; the direction of angulation is best termed fracture *apex* (anterior, posterior, etc.). Alternatively, the direction of angular displacement of the distal fracture fragment

may be specified.

   a. *Varus* angulation: The distal part of the distal fragment points toward the body midline (the apex points away).

   b. *Valgus* angulation: The opposite of varus; the apex is directed toward the body midline.

7. *Rotation:* Both proximal and distal joints must be included on the same film for proper evaluation.

8. Additional definitions:

   a. *Chip* fracture: An isolated bone fragment; this is to be distinguished from an *avulsion,* a fragment that is separated by traction from an attached tendon or ligament.

   b. *Dislocation:* Complete loss of articular contact in a joint; *subluxation* is a partial loss; *diastasis* is separation of a slightly movable joint.

   c. *Stress* fracture: Occurs when abnormal stress, often in the form of frequent repetitions of normal stress, is placed on normal bone. *Pathologic* fractures occur when normal stress is placed on bone that is abnormal owing to tumor, metabolic bone disease, or other disease.

C. Treated fractures: Healing and complications.

1. Same descriptive terms are used and healing is evaluated.

2. Healing is affected by many factors: Age of patient, amount of local bone and soft tissue trauma, percentage of bone loss, location of the fracture, degree of immobilization, presence of infection, local malignancy, radiation necrosis, avascular necrosis, intra-articular extension, steroid treatment, and multiple systemic factors. Therefore, *delayed union* is a *clinical* rather than a radiographic diagnosis.

3. *Nonunion,* on the other hand, is a *radiographic* diagnosis: one sees no bridging bone, and the fracture fragment ends are rounded and sclerotic. Nonunions may be *hypertrophic* or *atrophic.*
4. Fracture complications to watch for:
   a. Avascular necrosis: Especially in subcapital femoral, intra-articular femoral condylar, and waist of proximal pole scaphoid fractures. Watch for increased density as an early sign of avascular necrosis (AVN).
   b. Gas gangrene and osteomyelitis (especially pin tract osteomyelitis, manifested as a dense ring sequestrum).
   c. Hardware failure (described more completely in XIV, below).
   d. Reflex sympathetic dystrophy: Severe regional osteoporosis accompanied by soft tissue trophic changes.
   e. Malunion: Displacement may remodel completely. Limb length discrepancy up to 2 cm can be compensated. Anteroposterior (AP) angulation may be compensated in a hinge joint, but varus or valgus angulation is not compensated and the fracture remodels poorly. Rotational malunion does not remodel at all and is only partially compensated for in a ball and socket joint; it is totally uncompensated in a hinge joint.
   f. Remember to simply describe the findings rather than using qualitative terms such as "good" or "acceptable". Different criteria apply to fractures at different sites.
D. Fractures in childhood are unique in three ways.
   1. The bones are more porous, often resulting in incomplete fractures.
   2. There is a greater potential for remodeling malaligned fractures depending on:
      a. Number of years of growth left.

  b. Fracture near the growing end of the long bone.
  c. Whether the angular deformity is in the plane of movement of the adjacent joint.
3. The epiphyseal plate is the weakest, and one of the most easily fractured, sites in the long bone. The Salter-Harris classification allows easy radiographic classification, is a reasonable indicator of prognosis, and is a guide to treatment.
  a. Salter 1: Fracture through the plate itself, often unrecognized and having minimal displacement.
  b. Salter 2: Fracture through the plate and extending through the metaphysis. This is the most common of these fractures, and prognosis for healing without deformity is good.
  c. Salter 3: Fracture through the plate and extending through the epiphysis.
  d. Salter 4: Fracture through the epiphysis, crossing the plate, and extending through the metaphysis.
  e. Salter 5: Crush injury to the epiphyseal plate, often unrecognized or misdiagnosed as a Salter 1 injury. Salter 4 and 5 injuries are relatively rare but have very high complication rates, with partial premature epiphyseal plate closure and resultant deformity.
E. The role of other imaging modalities:
  1. Bone scan can detect occult fractures with nearly 100% sensitivity (exceptions may be the very osteoporotic patient and the patient on chronic steroid therapy). Usually they are used to detect acute and occult subcapital femoral neck, waist of scaphoid, and various stress fractures.
  2. Gallium or leukocyte scans may identify chronic osteomyelitis, which may show little

specific change on plain films.
3. Polytomography may detect subtle fractures and is used most commonly in the upper cervical spine. Computed tomography (CT) may also be useful in the individual case.

II. HAND TRAUMA

> **Key Concepts:** Avulsion injuries are easily missed and are functionally important. Watch especially for trauma to the dorsal and ventral aspects of the distal and middle phalanges and to the ulnar aspect of the base of the proximal phalanx of the thumb.

A. Tuft fracture.
B. Shaft fracture: Usually angulated; degree is underestimated except on lateral film. Rotational malalignment is an important parameter in functional impairment and is often detected only clinically.
C. Avulsion fracture: Usually dorsal or volar aspect (Fig 3–1).
   1. Baseball (mallet) finger (Fig 3–2):

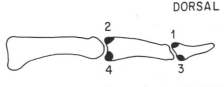

FIG 3–1.
Usual sites of phalangeal avulsion fractures: **1**, baseball finger; **2**, boutonnière; **3**, flexor digitorum profundus; **4**, volar plate.

DIP
FLEXED

FIG 3–2.
Baseball (mallet) finger: DIP flexion, with or without avulsion of dorsal aspect of the base of the distal phalanx.

  a. Due to flexion of a forcibly extended finger.
  b. Site of insertion of the common tendon of the extensor mechanism.
  c. Therefore results in either a tendon injury or a dorsal intra-articular avulsion fracture at the base of the distal phalanx.
  d. Clinically, has flexion and volar subluxation of the distal interphalangeal joint (DIP), sometimes with hyperextension of the proximal interphalangeal joint (PIP).
  e. Isolated flexion (the inability to extend the DIP joint while extending the PIP) should raise suspicion of this injury since otherwise this is difficult to do.
  2. Boutonnière (buttonhole) deformity.
  a. Due to PIP flexion with DIP extension and rupture of the middle slip of the extensor mechanism as it passes over the PIP.
  b. The extensor mechanism consists of one middle slip and two lateral slips at the PIP. The middle slip inserts at the base of the middle phalanx, and the two lateral slips join distally to form the common extensor tendon, inserting at the base of the distal phalanx (Fig 3–3).
  c. With rupture of the middle slip, the lateral slips migrate more volarly, forming a buttonhole through which the PIP flexes.

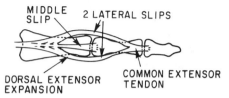

MIDDLE SLIP   2 LATERAL SLIPS

DORSAL EXTENSOR
EXPANSION

COMMON EXTENSOR
TENDON

FIG 3–3.
AP view of the extensor mechanism, with the middle slip shown inserting at the base of the middle phalanx and the two lateral slips joining to insert at the base of the distal phalanx.

The DIP is pulled into hyperextension (Fig 3–4).
  d. Avulsion of the dorsum of the base of the middle phalanx is unusual but may occur.
  e. Boutonnière deformity often occurs some time after the initial injury, as the lateral slips may migrate volarly slowly to form the buttonhole.
  f. It is important to diagnose the injury early

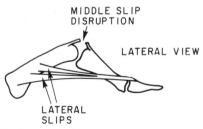

MIDDLE SLIP
DISRUPTION

LATERAL VIEW

LATERAL
SLIPS

FIG 3–4.
Lateral view of a boutonnière deformity, with PIP flexion and DIP extension, with or without avulsion of the dorsal aspect of the base of the middle phalanx.

on the basis of area of tenderness, soft tissue swelling, and possible avulsion fragment to facilitate early splinting and healing without development of the deformity.

3. Avulsion of a volar fragment of the proximal base of the distal phalanx (Fig 3–5).
   a. Forced hyperextension of a flexed finger avulses the flexor digitorum profundus from the volar aspect of the distal phalanx.
   b. May be either a pure tendon injury (which requires a stress film for diagnosis) or may produce an intra-articular avulsion fracture.
   c. If a fragment is present, it may be retracted proximally to the volar side of the middle phalanx.
   d. Clinically, the DIP cannot be flexed.
4. Volar plate fracture.
   a. Volar plate: A fibrocartilaginous structure crossing the metacarpophalangeal (MCP) and PIP joints that has a weak proximal attachment but stronger distal attachment at the base of the middle phalanx volarly.

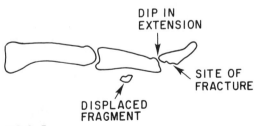

FIG 3–5.
Avulsion of flexor digitorum profundus, with DIP hyperextension, with or without avulsed fragment.

  b. Hyperextension will avulse this distal
     attachment, yielding a fragment from the
     volar side of the base of the middle
     phalanx (Fig 3–6).
  c. May be seen in conjunction with mallet
     finger.
5. Collateral (radial and ulnar) ligaments are
   found at all MCP, PIP, and DIP joints.
   a. May be ruptured or may avulse
      fragments.
   b. Most common is avulsion of the ulnar
      collateral ligament of the MCP of the
      thumb—"gamekeepers thumb"—due to a
      valgus injury (today, commonly due to ski
      pole injury), may give an intra-articular
      avulsion fragment at the base of the
      proximal phalanx (ulnar side). If not,
      stress views are required.
   c. Stress views of the thumb must be truly
      AP (requiring full pronation of the wrist)
      and should be compared with the
      opposite, normal side.
D. Boxer's fracture.
   1. Usually fifth metacarpal.
   2. Apex dorsal angulation is best appreciated on
      the lateral view.
   3. Volar comminution makes a stable reduction
      difficult, so watch for reestablishment of the
      apex dorsal angulation in follow-up films.
E. Thumb: NB: standard *hand* films *do not* give true

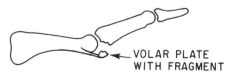

FIG 3–6.
Volar plate fracture with PIP hyperextension.

AP and lateral views of the *thumb* and, so, are inadequate for evaluation of thumb trauma. Standard hand films demonstrate the thumb only in varying degrees of obliquity, often giving a false appearance of subluxation at the carpometacarpal (CMC) and MCP joints. Standard thumb views are obtained as follows:

1. AP: Hand fully pronated with the dorsal surface of the thumb held against the film.
2. Lateral: Hand pronated 15 degrees.

F. Fracture of the first metacarpal.

1. It is important to differentiate between intra- and extra-articular fractures, since the former may require open reduction and the latter may be treated closed.
2. One third of first metacarpal (MC) fractures are *Bennett* variety—a fracture dislocation with an oblique intra-articular fracture at the base of the MC and dorsal dislocation or subluxation of the MC (the smaller fragment retains its articulation with the trapezium). These are usually treated by open reduction and internal fixation.
3. Far fewer are *Rolando* fractures—comminuted Bennett's fracture with dorsal subluxation and a separate dorsal fragment. These are usually treated closed with casting since pin fixation is not successful in the presence of much comminution.

G. Dislocations.

1. Phalanx: Usually dorsal dislocation, produced by hyperextension.
   a. Volar plate disruption with or without associated volar avulsion.
   b. May be unreducible owing to joint capsule or volar plate interposition. The only radiographic sign of this complication is an interposed sesamoid at the MCP.
2. CMC[1]:
   a. Uncommon site of dislocation.

    b. Fifty percent involve the fifth MC.

    c. Twenty-five percent involve the second MC.

    d. Eighty percent are multiple, usually involving the fifth MC plus another.

    e. Two thirds dislocate dorsally.

    f. The posteroanterior (PA) view is very useful: The palm must be *flat* on the cassette; flexion of the fingers gives a false appearance of overlap at the CMC joints.

    g. On the PA view, the CMC joint spaces are of an even width, the opposing carpal and MC joint surfaces are parallel, and there is a zig-zag pattern of articulation (parallel M; Fig 3–7).

    h. First MC articulates only with the trapezium.

    i. Second MC articulates with the trapezoid. A styloid process is located at the base of the second MC on the ulnar side, which articulates with the trapezoid, capitate, and third MC in a lock-and-key configuration.

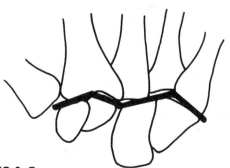

FIG 3–7.
Articulation of the carpometacarpal joints.

  j. Third MC articulates with the capitate.
  k. Fourth MC articulates with the hamate (radial side).
  l. Fifth MC articulates with the hamate (sloping ulnar side).
  m. A fracture at the base of an MC or involving the distal carpal row should stimulate a search for an occult dislocation.

III. WRIST TRAUMA

> **Key Concepts:** A fall on an outstretched hand can result in different injuries, depending on age of patient: fracture of both bones of the forearm in a young child; Salter II distal radial fracture in an older child; scaphoid fracture in a teenager or young adult; and Colles' fracture in an older adult. Wrist fractures often are not displaced and, therefore, are occult. The PA film must always be examined carefully for continuity of the three arcs. The lateral film is essential for diagnosis of carpal dislocations and instability patterns. The coaxial arrangement of the radius-lunate-capitate must always be demonstrated.

A. Distal forearm:
  1. PA film: Normal anatomy.
    a. Distal radial articular surface tilts 17 degrees toward the ulna (Fig 3–8, A).
    b. Distal radial epiphyseal line may remain a linear density, with a spurlike projection laterally simulating an avulsion.
    c. Medially, the radius articulates with the head of the ulna at the ulnar notch. The head of the ulna is usually 1 to 2 mm shorter than the radius and either touches or slightly overlaps the radius at the distal radioulnar joint.

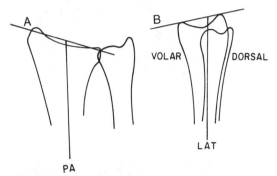

FIG 3–8.
A, normal ulnar tilt of the distal radial articular surface on the PA film. **B,** normal volar tilt of the distal radial articular surface on the lateral film. Note that the distal ulna is slightly shorter and posteriorly placed with respect to the radial articular surface on a well-positioned lateral film.

2. Lateral film: Normal anatomy.
   a. Pronator quadratus bulge may be a soft tissue clue to injury.
   b. Distal radial articular surface is angled 10 to 15 degrees volarly (Fig 3–8, B).
   c. On a true lateral film, the dorsal surface of the distal ulna lies 1 to 3 mm posterior to the dorsal surface of the distal radius.
3. Distal forearm abnormalities: Ulnar variants.
   a. Ulnar-plus variant: An unusually long ulna bears more of the axial load; this is associated with lunate-triquetral ligament damage.
   b. Ulnar-minus variant: An unusually short ulna bears less of the axial load; this is associated with lunate osteomalacia (Keinbock's disease).

4. Distal radioulnar dislocation.
   a. Suggested by an abnormal position of the head of the ulna in the ulnar notch.
   b. Diagnosis is best made by a single CT axial image through the distal radioulnar notch.
   c. May be seen as an isolated injury or may be part of a more complex injury (Colles' fracture or Galleazzi's fracture-dislocation).
5. Fractures of the distal forearm:
   a. The most common injuries of the skeletal system.
   b. Most result from a fall on the outstretched hand.
   c. Age alone is a very good predictor of the injury:
      (1) 4–10 years: Transverse fracture of the metaphyses of the distal radius and ulna.
      (2) 11–16 years: Salter fracture of the distal radial epiphysis.
      (3) 17–40 years: Fracture of the scaphoid.
      (4) >40 years: Colles' fracture.
   d. Transverse metaphyseal fractures seen in children:
      (1) May be complete or incomplete.
      (2) When incomplete, may be greenstick or torus (buckle).
   e. Salter fractures of the distal radial epiphysis.
      (1) Salter 2 most common.
      (2) Usually displaced dorsally, so often are seen *only* on the lateral film.
   f. Distal radial fractures in the adult.
      (1) Colles':
         a Most common.
         b The injury is more common in females than in males owing to the higher incidence of senile osteoporosis in women.

   c Associated with fractures of proximal humerus and hip due to falls in osteoporotic patients.

   d Apex volar angulation with dorsal impaction.

   e May or may not have an intra-articular component.

   f May or may not have an associated ulnar styloid fracture.

   g Complications visible on postreduction films:

    i Apex volar angulation and consequent loss of the normal volar tilt of the radial articular surface.

    ii Loss of radial length due to impaction and consequent ulnar plus variant

    iii Intra-articular diastasis with loss of the normal ulnar tilt of the radial articular surface.

 (2) Smith: Reverse Colles', with apex dorsal angulation.

 (3) Barton's: Intra-articular fracture of the dorsal lip of the radius; the carpus follows the dorsal fragment. This is an unstable fracture that often requires internal fixation or even placement of an external fixator.

 (4) Reverse Barton's: Intra-articular fracture of the volar lip of the radius.

 (5) Hutchinson's or chauffeur's: Intra-articular fracture of the radial styloid process. (Do not be fooled by the normal notch at the distal radial epiphysis.)

B. Carpus.

 1. Frequency of carpal injuries.[2]

  a. Forearm injuries are 10 times as frequent as carpal injuries.

   b. Carpal injuries are rare in patients under
      12.
   c. Of carpal injuries, 60% to 70% are
      scaphoid fractures; 10%, dislocations and
      fracture-dislocations; 10%, dorsal chip
      fractures (usually triquetrium); and 10%,
      others.
2. Normal anatomy of the carpus (PA).
   a. Proximal and distal rows bridged by
      scaphoid.
   b. Width of intercarpal joints is uniform,
      about 2 mm.
   c. Three parallel arcs have been described by
      Gilula[3] to aid in evaluation (Fig 3–9):
      (1) Along the proximal articular margins of
          the proximal carpal row.

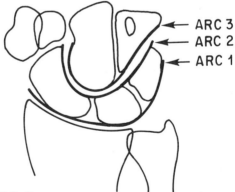

← ARC 3
← ARC 2
← ARC 1

FIG 3–9.
Normal PA carpal view demonstrating particularly the
normal shapes of the scaphoid and lunate as well as
the three continuous arcs of the proximal and distal
carpal rows. Continuity of these arcs assures carpal row
integrity.

    (2) Along the distal articular margins of the proximal carpal row.

    (3) Along the proximal articular margins of the distal carpal row.

d. Fifty to 70% of the lunate articulates with the radius.

e. The lunate is trapezoidal on the PA film. If the shape is triangular, the lunate is rotated abnormally. (This may be seen as an artifact with the hand held in flexion or extension, however.)

f. The hook of the hamate is normally seen *en face* in the PA film overlying the hamate. Its absence indicates fracture, hypoplasia, or an ununited os hamuli proprium.

g. Radial or ulnar deviation can alter the appearance of the PA carpus.

    (1) With radial deviation, only 25% of the lunate articulates with the radius, the scaphoid is foreshortened and distorted. Patients with painful wrists normally hold the hand in radial deviation, so scaphoid fractures may easily be missed!

    (2) With ulnar deviation, 100% of the lunate articulates with the radius and the scaphoid appears elongated.

h. Normal variants of carpus, PA view:

    (1) Multiple accessory ossicles; the most common may be adjacent to the ulnar styloid process and may be indistinguishable from a previous ununited fracture. Others are outlined in Theodore E. Keats' *Atlas of Normal Roentgen Variants That May Simulate Disease.*

    (2) Carpal fusions (most commonly lunate-triquetrum; these may be only partially fused and may simulate fracture).

(3) Bipartite scaphoid; these may be old nonunions.
3. Normal anatomy of carpus, lateral view (Fig 3–10).
    a. Confusing overlap, but the coaxial relationship between the articular surfaces of the radius, lunate, capitate, and third metacarpal *must* be picked out. An exact linear relationship here is uncommon, but a coaxial relationship is essential.
    b. The scaphoid arches out on the palmar side.
    c. In a true lateral view, the dorsal carpal surface is formed by the triquetrum.
4. Normal anatomy of the carpus, carpal tunnel view:
    a. The pisiform and hook of the hamate are projected on the ulnar side; the scaphoid and trapezium (with a small hook) are projected on the radial side. Either the

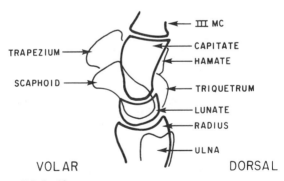

FIG 3–10.
Normal lateral carpal view, in which one must pick out the articulating distal radius, lunate, capitate, and third metacarpal (*bold outline*).

lunate or the capitate is seen centrally, depending on how the beam is angled.

b. The view may be useful to search for calcifications or osteophytes in a patient with carpal tunnel syndrome. It has been suggested to diagnose fracture of the hook of hamate, but since these fractures are usually at the base of the hook and the patients usually cannot hyperextend the hand for the carpal tunnel view, the diagnosis may be difficult in injured patients. A supinated oblique view, limited CT, or lateral tomograms seem to be more useful in the diagnosis.

C. Carpal fractures.

1. Scaphoid fractures.

a. Seventy percent are at the waist and nondisplaced; if clinically suspected but not seen, films may be repeated in 7 to 10 days, when sclerosis or resorption about the fracture line may be seen.

b. If it is essential to make the diagnosis acutely, tomograms or bone scan may help.

c. The blood supply to the distal pole is separate, so distal pole fractures heal quickly.

d. The blood supply to the proximal pole enters at the waist, so it may be cut off by a fracture through the waist of the scaphoid. Such fractures are at risk for delayed union, nonunion, and avascular necrosis of the proximal pole fracture fragment.

e. A waist of scaphoid fracture may take up to 2 years to unite. It may in fact develop radiographic signs of nonunion (rounded, sclerotic fracture fragment edges) yet go on to unite after several months; 90% of scaphoid fractures unite eventually.

2. Lunate fractures: It is unusual to see an acute fracture, but the lunate is vulnerable to avascular necrosis (Keinbock's disease or lunate malacia), so collapse and increased density may be seen. This is felt to be secondary to trauma (though the incident may not be recalled) and has been associated with the ulnar minus variant.

3. Triquetrum fractures: Dorsal chip fractures are usually triquetral and are seen only on the lateral film.

4. Capitate fractures: Isolated fracture is rare, but as a part of a complex fracture-dislocation, a transverse fracture of the capitate may be seen; in this case, the proximal fragment frequently rotates 180 degrees.

5. Hamate fractures.
   a. Hook of the hamate fracture is best detected on PA films (by its absence) and confirmed by a carpal tunnel view, CT, or tomography.
   b. Proximal pole fractures may be seen as part of a perilunate or lunate fracture-dislocation complex.

D. Carpal dislocations.
   1. Most fracture-dislocations of the carpus fall within the vulnerable zone outlined by Yeager[4]: The inner arc roughly outlines the disrupted ligaments around the lunate, as seen in a pure dislocation, while the outer arc outlines the fractures commonly associated with carpal fracture-dislocation (radial styloid, waist of scaphoid, proximal capitate, base of hamate, lunar surface of triquetrium, and ulnar styloid).
   2. With progression from the radial to the ulnar side of the arch, the severity of the injury increases (i.e., a scapholunate dissociation is less severe than a perilunate dislocation,

which in turn is less severe than a lunate dislocation).

3. With progression from the radial to ulnar side, the frequency of the injury decreases (i.e., a scapholunate dissociation may be a less severe injury than a perilunate dislocation, but it is also much more common; perilunate dislocations are more common than lunate dislocations).
4. Fracture dislocations are more common than pure dislocation, so look carefully for these fractures.
5. Radiographic analysis: On the PA film, Gilula's first and second arcs are disrupted; on the lateral film, the coaxial radius-lunate-capitate arrangement is disrupted (Fig 3–11,A).
6. Perilunate dislocation: The capitate articular surface is dislocated from the lunate (almost invariably dorsally); the lunate maintains its normal articulation with the radius (Fig 3–11,B).
7. Midcarpal dislocation: The lunate tilts volarly but is not dislocated from the radius. The capitate is dislocated from the lunate but is not as dorsally placed as in a routine perilunate dislocation. This may be a carpus in transition from a perilunate to a lunate dislocation (Fig 3–11,C).
8. Lunate dislocation: The lunate has lost its articulation with both the capitate and radius and is displaced volarly with 90 degrees' rotation. The capitate remains aligned with the radius but sinks proximally (Fig 3–11,D).

E. Carpal instabilities.

1. Ligamentous injury can give instability patterns without frank dislocations.

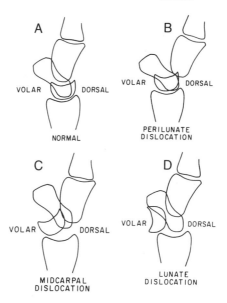

FIG 3–11.

Carpal dislocation patterns as seen on the lateral film. The progression from normal **(A)** to perilunate (loss of articulation of the capitate with the lunate) **(B)** through midcarpal (dislocation of the capitate from the lunate, and subluxation but not dislocation of the lunate from the radius) **(C)** to lunate **(D)** (dislocation of the capitate from the lunate and of the lunate from the radius) demonstrates increasing severity of injury as well as the possible sequence of injury patterns in a single person. A patient who presents initially with a perilunate dislocation may go on to a midcarpal dislocation, and convert to a lunate dislocation if there is sufficient ligamentous disruption.

2. Radiographic examination.
   a. PA: Look especially for a diastasis at the scapholunate joint (normal scapholunate gap is 2 mm) and lunate tilt.
   b. Lateral: Note the previously discussed coaxial relationships. Also, evaluate the lunate-capitate angle (normally <20 degrees) and the scapholunate angle (normally 30 to 60 degrees; under 30 and over 80 degrees definitely are abnormal; see Fig 3–12,A).
3. Fluoroscopy: Occasionally wrist instability patterns are transient and can be seen only under fluoroscopy. This is especially true in patients who have midcarpal pain, normal plain films, and audible clicks.
4. Standard fluoroscopy (with videotaping) includes:
   a. PA: Radial to ulnar deviation with fist clenching.
   b. Lateral: Dorsiflexion and palmar flexion.
   c. AP: Fist clenching.
   d. Anything else to elicit and define a click.
5. Scapholunate dissociation may be accentuated by AP filming with a clenched hand and radial or ulnar deviation, which drives the capitate proximally between the scaphoid and lunate, widening that gap.
6. Capitate-lunate instability may be seen best when pressure is applied to the scaphoid tuberosity and longitudinal traction is maintained while the hand is flexed (both PA and lateral projections).
7. Triquetrum-hamate instability: During radial-to-ulnar deviation, the proximal carpal row snaps from palmar flexion to dorsiflexion

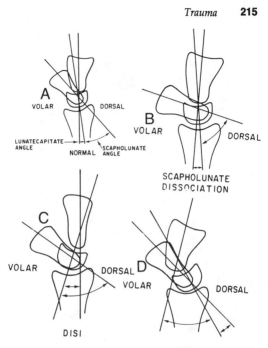

FIG 3–12.
Carpal instability patterns. **A**, normal, with lunate-capitate angle less than 20 degrees and scapholunate angle of 30 to 60 degrees. **B**, scapholunate dissociation with rotary subluxation of the scaphoid (normal lunate-capitate angle, abnormally large scapholunate angle). **C**, dorsiflexion carpal instability pattern, with increased lunate-capitate angle, increased scapholunate angle, and dorsiflexion of the lunate. **D**, volar flexion carpal instability, with increased lunate-capitate angle, often decreased scapholunate angle, and volarflexion of the lunate.

once ulnar deviation is obtained.

8. Scapholunate dissociation. Gap at scapholunate joint is greater than 2 mm. (Fist clenching may help, see discussion above). Patient may or may not have scaphoid rotation, manifested on PA film by scaphoid foreshortening and "ring" sign and on lateral by increase in scapholunate angle without abnormal lunate flexion (Fig 3–12,B).

9. Dorsiflexion carpal instability (also termed dorsal intercalated segment instability—DISI; Fig 3–12,C).

   a. Increased dorsiflexion of lunate, with consequent increase in lunate-capitate angle on lateral.

   b. Palmar flexion of scaphoid, with consequent increase in scapholunate angle on lateral.

   c. May or may not have scapholunate dissociation.

10. Volar flexion carpal instability (also termed volar intercalated segment instability—VISI; Fig 3–12,D).

   a. Volar flexion of lunate and dorsiflexion of capitate, giving a zig-zag deformity and abnormal lunate-capitate angle on lateral.

   b. Often, but not invariably, there is a decrease in scapholunate angle.

   c. Rare instability pattern compared to DISI if secondary to trauma but relatively common if secondary to ligamentous instability in rheumatoid arthritis.

11. Ulnar translocation: Entire carpus translocates ulnarly from the radial styloid process, usually secondary to an inflammatory process such as rheumatoid arthritis (RA).

12. Dorsal carpal subluxation: Dorsal subluxation relative to the distal radius, often secondary to a residual fracture deformity such as Colles' fracture with apex volar angulation.

F. Wrist arthrograms.
    1. Always examine with fluoroscopy first to look for instability.
    2. Objective is usually to rule out disrupted triangular fibrocartilage complex (TFCC) or interosseous ligaments.
    3. Dorsal approach, wrist held in slight flexion, needle angulated proximally to avoid the lip of the radius; 25-gauge, $1\frac{1}{2}$-inch needle placed between the scaphoid and radius (and specifically *not* at the scapholunate junction). Two milliliters of contrast (60% in 10:1 mixture with epinephrine, 1:1000) injected using digital subtraction technique. Follow this with PA, lateral, and oblique films as well as a postexercise series.[5]
    4. Contrast normally is retained in the radiocarpal compartment. Communication between the triquetrum and pisiform is a normal variant.
    5. Contrast entering the distal radioulnar joint indicates TFCC disruption.
    6. Contrast entering the midcarpal joint indicates disruption of one of the following interosseous ligaments: scapholunate, lunate-triquetral, triquetral-hamate, scapho-trapezium. Identification of the exact site of disruption is important if surgery is contemplated; this can usually be accomplished by a frame-by-frame analysis of the digital portion of the study but often is missed otherwise.
    7. Degeneration of these ligaments and the TFCC may occur normally in the older age group. It is also a common finding in patients with erosive arthropathies.
    8. If the radiocarpal injection is normal, a distal radioulnar joint injection should be performed as well as a midcarpal injection (at the scaphoid-capitate joint), in order to avoid

missing a ball-valve type of ligamentous perforation.

IV. ELBOW TRAUMA

> **Key Concepts:** For elbow trauma, must evaluate: fat pads (provide a hint that an occult fracture is likely to be present); ossification centers, especially medial epicondyle (knowing the epiphyseal maturation sequence is useful, especially to avoid missing a medial epicondylar avulsion); anterior humeral line (diagnoses occult supracondylar fracture); radiocapitellar line (diagnoses occult radial head dislocation): if there is a dislocation, look for fracture.

A. Normal anatomy.
  1. AP film:
    a. Radius articulates with capitellum.
    b. Ulna articulates with trochlea.
    c. Normal carrying angle 165 degrees (on AP, apex valgus; functionally, this displaces the hand away from the thigh).
    d. Bowman's angle describes the normal cubitus valgus in a child and is used to evaluate for abnormal valgus or varus in the presence of a supracondylar fracture. The angle is formed by a line bisecting the humeral shaft and a second line along the metaphysis of the medial epicondyle. This angle decreases with varus angulation. The abnormal and normal elbows are compared; 5 degrees is considered a significant difference in angulation (Fig 3–13).
  2. Lateral film.
    a. The condyles of the humerus are normally anteriorly placed with respect to the

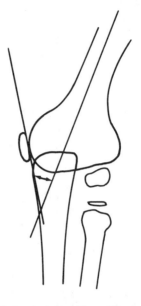

FIG 3–13.
Bowman's angle, decreased in cubitus varus resulting from malalignment of a supracondylar fracture.

humeral shaft, described by the *anterior humeral* line: on a true lateral film, a line drawn along the humeral cortex intersects the capitellum in its *middle third* (Fig 3–14).

b. Radiocapitellar line: With the elbow in *any* position, a line bisecting the proximal radial shaft intersects the capitellum. If it does not, there is a radial head dislocation (Fig 3–15).

c. Fat pads: Seen on lateral film, flexed 90 degrees, the anterior fat pad is normally seen in the coronoid fossa as a straight

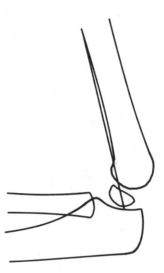

FIG 3–14.
Anterior humeral lines on a true lateral film must intersect the middle third of the capitellum. In a supracondylar fracture, the anterior humeral line usually intersects the anterior third of the capitellum (or none) since the distal fragment is displaced or angulated posteriorly. An abnormal anterior humeral line and abnormal fat pad may be the only clues to a supracondylar fracture since the fracture line itself often is occult.

(not convex) lucency. In the absence of effusion, the posterior fat pad is not seen; with an effusion, the posterior fat pad is a convex lucency in the olecranon fossa of the humerus. With a small effusion, an abnormal (convex) anterior fat pad may be seen without a posterior fat pad. In the

presence of trauma and abnormal fat
pads, a fracture must be ruled out.
Multiple views may be necessary.
B. Normal variants.
1. There is a normal chevron appearance of the
trabeculae seen in the lateral humerus in the
supracondylar region.
2. When the radial tuberosity is seen *en face*, it
may present as a well-circumscribed lucency,
possibly simulating neoplasm.
3. Supratrochlear ossicle: A normal variant

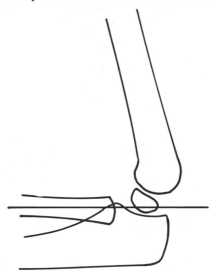

FIG 3–15.
Radiocapitellar line: A line bisecting the proximal shaft
of the radius must intersect the capitellum in *any* position of the elbow; an abnormal radiocapitellar line is
the best indicator of subtle radial head dislocation.

found in the trochlear fossa, to be
distinguished from a chip fracture.
4. Supracondylar process: A hooklike process
arising from the anterior humeral cortex in
the supracondylar region that may
occasionally fracture and cause a median
nerve injury.
C. Epiphyseal maturation sequence.
1. Capitellum: 1 year.
2. Radial head: 3 to 6 years.
3. Medial epicondyle: 5 to 7 years; last to fuse.
4. Trochlea: 9 to 10 years; ossifies in multiple
centers, so may appear fragmented.
5. Olecranon: 6 to 10 years; two separate centers
of ossification appear markedly separated
from the metaphysis.
6. Lateral epicondyle: 9 to 13 years; sliver shape
initially, located very lateral to the
metaphysis, so may mimic on avulsion.
7. The timing is not as important as the
sequence. Specifically, note that the trochlea
never ossifies before the medial epicondyle.
Since the flexor muscles of the forearm arise
from the medial epicondyle, with avulsion
the medial epicondyle ossification center is
pulled inferiorly and may become trapped in
the joint. In a patient who is too young to
have an ossified trochlea, this avulsed
fragment may be mistaken for the trochlea;
thus, the diagnosis is easily missed. NB: The
presence of a "trochlea" in the absence of a
medial epicondylar ossification center is
diagnostic of an avulsed and trapped medial
epicondyle.
D. Elbow fracture in children.[6]
1. Supracondylar fractures 60%: Almost always
have posterior displacement of the condyles
and an abnormal anterior humeral line. (See
discussion above.)

       a. May be occult except for abnormal fat pads and the anterior humeral line extending abnormally through the anterior portion of the capitellum.

       b. May result in abnormal cubitus varus or valgus; Bowman's angle and the opposite elbow are useful for the evaluation. (See discussion above.)

   2. Lateral condylar 15%.

       a. Usually Salter 4, involving the capitellum.

       b. The extensor muscle mass attaches to the metaphyseal fragment and displaces it distally and posteriorly. Internal fixation is required.

   3. Medial epicondyle avulsion 10%: May be trapped in the joint; often unrecognized, resulting in severe late disability. (See discussion of epiphyseal maturation sequence, C, above.)

E. Elbow fractures in adults.

   1. Radial head fractures.

       a. Half of all adult elbow fractures.

       b. Most common cause of positive fat pad signs in adults.

       c. Half are undisplaced, so several views may be required.

       d. Essex-Lopresti fracture: a severely comminuted radial head fracture associated with subluxation of the distal radioulnar joint.

   2. Olecranon fracture: If distal to the triceps insertion, the fracture is widely displaced.

   3. Transcondylar fracture: Generally in older patients with osteoporosis.

   4. Intracondylar T or Y: Most distal humeral fractures in adults are this type or a more comminuted variant. The vertical component is usually at the trochlear ridge. Check for diastasis of the condyles at this site.

F. Dislocated elbow.

1. Eighty to 90% are posterolateral.
2. Often have associated coronoid process or radial head fractures, which may become trapped in the joint.
3. An isolated dislocated radial head is rare in adults. An ulnar fracture (Monteggia's) must be excluded.
4. Myositis ossificans is a relatively frequent complication, associated most often with fracture-dislocations. It is usually located ventrally, in the brachialis muscle. There is an increased incidence of myositis with delay in treatment.
5. Congenital elbow dislocation: There is associated radial head overgrowth and an abnormally shaped radial head articular surface, so it is easily differentiated from traumatic dislocations.
6. Nursemaid's elbow: Traumatic subluxation of the radial head from the annular ligament by sudden extension of the elbow. The radiograph is normal with the hand in pronation. It often reduces spontaneously when the forearm is supinated for the AP film.
7. Separation of the distal humerus in children initially may look like a dislocation. The radial head, however, maintains its normal relationship with the capitellum, and the capitellum is no longer aligned with the humerus. The fragment tends to be displaced medially rather than posterolaterally as in most dislocations of the elbow.

V. SHOULDER TRAUMA

Key Concepts: Axillary lateral view is essential in evaluating shoulder trauma and may be obtained using only minimal abduction in an injured shoulder. Anterior dislocation of humeral head is much more common, but posterior dislocation is more often missed.

A. Sternoclavicular (SC) joint.
   1. Fifty percent of the clavicular articular surface extends above the manubrium, so the normal articulation has an unusual (though bilaterally symmetric) appearance.
   2. A "bump" noted clinically at the SC joint most likely represents a dislocation or infection.
   3. Either entity is best diagnosed by a limited CT scan tailored to the area. This allows definition of both soft tissue and bone detail as well as evaluation of alignment. Only two or three axial cuts are usually required, which certainly is cost effective compared to polytomography.
   4. If plain films are requested, an AP with 40 degrees' cephalic angulation is likely to be the most useful.
   5. The medial clavicular epiphysis ossifies late (18 to 20 years) and fuses late (about 25 years). As this is the age range in which most SC dislocations are seen, it is felt that many of them are actually Salter 1 or 2 fractures.
   6. Dislocation may be anterior or posterior (the clavicle tends to move slightly superiorly in either case).
   7. Anterior is more common than posterior.
   8. A posterior dislocation has the potential to compress the great vessels or trachea. Surgical reduction may be required, so accurate diagnosis is mandatory. It should be

noted that the surgical complication rate in this injury is also high.

B. Clavicle.
  1. Rhomboid fossa: Concave and irregular fossa located at the medial clavicle inferiorly; site of attachment of costoclavicular ligaments; may be asymmetric and may appear quite destructive.
  2. Fractures.
      a. Eighty percent involve the middle third. Callus formation may be exuberant.
      b. Fifteen percent involve the distal third. In these cases, the integrity of the coracoclavicular (CC) ligament must be evaluated, just as in an acromioclavicular (AC) separation.
  3. Distal clavicular resorption: Differential diagnosis:
      a. RA.
      b. Hyperparathyroidism.
      c. Infection.
      d. Traumatic osteolysis: Fairly massive osteolysis following major or minor trauma which stabilizes by approximately 18 months and progresses to a reparative phase; associated with weight lifting.

C. AC joints.
  1. Evaluation requires single film demonstrating both AC joints, filmed with and without weights suspended from wrists (not grasped by hands) for stress.
  2. AC separation may involve the AC joint itself (and a tear of the AC ligament) as well as the CC ligaments (conoid and trapezoid).
      a. Type I: Sprain, with intact ligaments and normal radiograph.
      b. Type II: AC ligament disruption with widening of the AC joint; CC ligaments may be stretched slightly but are intact.

  c. Type III: Both AC and CC ligaments are disrupted, allowing widening of the AC joint and droop of the scapula ("elevation" of the clavicle relative to the acromion and coracoid processes).

  d. Type IV: Posterior dislocation of clavicle relative to acromion; usually seen only on axillary lateral film since often no AC widening is seen on AP film.

 3. The width of the AC joint is normally 3 to 5 mm but may occasionally be as wide as 8 mm. The normal difference between sides should be less than 2 to 3 mm.

 4. The width of the CC ligament is 11 to 13 mm. The difference between left and right side should be less than 5 mm.

 5. Rule of thumb: These numbers may be difficult to remember, but they translate roughly to a 50% side-to-side difference in width of either the AC or CC being significant.

 6. The coracoid apophysis may be avulsed as part of the type III separation.

D. Scapula.

 1. Fractures are uncommon and usually are not displaced significantly because of extensive muscular coverage.

 2. Ossification centers not to be mistaken for fracture: Coracoid, acromion (with occasional os acromiale), glenoid rim, and longer apophyses along the vertebral border and inferior angle.

E. Shoulder joint anatomy (Fig 3–16).

 1. Glenoid is shallow, deepened somewhat by the cartilaginous labrum, and maintained primarily by muscular forces.

 2. Glenoid is directed anteriorly, so AP film shows overlap of humeral head and glenoid. A 45-degree posterior oblique film shows the joint in tangent.

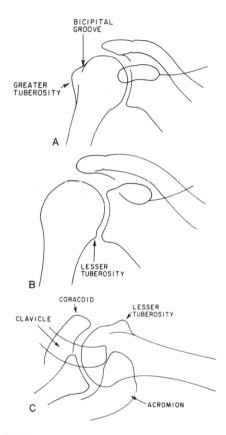

FIG 3–16.
Normal shoulder. **A,** external rotation. **B,** internal rotation.
C, axillary lateral.

3. In external rotation, the greater tuberosity is seen laterally and the lesser tuberosity is superimposed over the humeral head.
4. In internal rotation the head appears more rounded, the lesser tuberosity is seen medially, and the greater tuberosity is superimposed over the humeral head.
5. The axillary lateral is the only true lateral view of the shoulder joint. The coracoid is seen to be anterior and the acromion, posterior. If properly positioned, the AC joint is centered over the humeral head.

F. Shoulder dislocations.
   1. Common site of dislocation due to the shallowness of the articulation.
   2. Anterior dislocations (95%):
      a. Humeral head moves anteriorly and medially.
      b. The posterolateral portion of the humeral head impacts on the anteroinferior portion of the glenoid.
      c. This wedge-shaped impaction, called a Hill-Sachs defect, is best seen on an AP film with internal rotation. The defect may be difficult to see on an axillary lateral film and may be missed entirely on an external rotation film.
      d. The corresponding chip fracture of the anteroinferior rim of the glenoid is called the Bankart lesion. It may be seen on either the anterior or axillary views, but often not both. If only the labrum is involved, it is best detected by double contrast arthrography with CT.[7]
      e. The Hill-Sachs or Bankart lesion may be seen either acutely (30% to 40%) or after recurrent dislocation.
      f. Detection of these lesions is important since they may predispose to recurrent dislocation and instability.

      g. Incidence of recurrence is inversely related to the patient's age at the time of initial dislocation.

  3. Posterior dislocation 2% to 4%.

    a. Usually secondary to shock therapy or seizures.

    b. Directly posterior, without superior or inferior change in alignment.

    c. Fifty percent are not recognized initially for two reasons: The subtle signs are missed on the AP film, and axillary lateral films are not obtained. The abnormality is very obvious on axillary lateral film, so these should be obtained routinely.

    d. The humeral head is usually locked in internal rotation. This is the most reliable sign in the absence of an axillary lateral.

    e. Trough sign: A vertical density parallel to the articular surface corresponding to a "reverse Hill-Sachs" impaction on the anteromedial portion of the humeral head. A "reverse Bankart" chip fracture of the posterior glenoid may be seen.

    f. Rim sign: The humeral head often appears to be slightly displaced laterally, giving a "widened" joint space. This appearance is highly dependent on positioning of the film.

  4. Inferior dislocation (luxatio erecta): A rare dislocation in which the articular surface is dislocated entirely inferior to the glenoid with fixed abduction of the shaft.

G. Rotator cuff tear.

  1. Rotator cuff consists of the supraspinatus, infraspinatus, teres minor muscles (inserting on the greater tuberosity; external rotators), as well as the subscapularis muscle (inserting on the lesser tuberosity, internal rotator).

2. The supraspinatus muscle and tendon primarily occupy the space between the humeral head and acromion process. It is this tendon that is torn.

3. Rotator cuff tears may be acute or chronic and are common in patients with RA.

4. Secondary signs of a chronic tear.
   a. Elevation of the humerus with a decrease in the acromiohumeral distance due to attenuation of the supraspinatus muscle.
   b. Inferior surface of the acromion becomes convex and sclerotic owing to the mechanical apposition of the humeral head.
   c. In RA, the complication of medial impaction of the surgical neck of the humerus on the inferior glenoid may occur. This leads to a mechanical erosion and an occasional fracture of the surgical neck.

5. Diagnosis is made by arthrography. Contrast normally enters the axillary and subscapularis bursae. In the presence of a tear, contrast enters the subacromial bursa, which communicates with the subdeltoid bursa. Diagnosis may also be made either by ultrasound or magnetic resonance imaging (MRI).

H. Shoulder impingement syndrome[8]:
   1. Impingement of the greater tuberosity and soft tissues on the coracoacromial ligmentous arch, generally during abduction of the arm.
   2. Secondary signs include a spur arising from the anterior inferior aspect of the acromion process as well as flattening and sclerosis of the greater tuberosity.
   3. May be reproduced under fluoroscopy with

abduction, external rotation, or elevation of
the humerus, eliciting a sharp pain.
4. Seen in young patients as well as older
patients.
5. Presentation similar to that of a rotator cuff
tear.
6. May progress to tendinitis or rotator cuff tear.
I. Proximal humerus fractures.
1. Relatively common, especially in osteoporotic
bones.
2. Neer's four-segment classification helps
assess prognosis and guide treatment:
   a. Four segments are involved: Humeral
   head, humeral shaft, greater tuberosity,
   lesser tuberosity.
   b. When any of the segments is displaced
   more than 1 cm or angulated more than 45
   degrees, it is considered significantly
   displaced.
   c. If one segment is displaced, it is termed a
   two-part fracture; a three-part fracture has
   two displaced segments, etc.
   d. If none of the fragments is significantly
   displaced or angulated, it is a
   nondisplaced fracture, no matter how
   comminuted it is.
   e. Eighty percent of proximal humeral
   fractures are nondisplaced. Fragments are
   held together by joint capsule,
   periosteum, and rotator cuff muscles.
   f. An axillary lateral film *must* be obtained
   since apex anterior angulation is relatively
   common and is not evaluated by the other
   views.
3. Pseudosubluxation of the shoulder following
such a fracture is temporary and is due to
hemarthrosis and/or atony of the deltoid and
rotator cuff muscles.
J. Humeral shaft.

1. Deltoid tuberosity is a normal lateral cortical thickening extending approximately halfway down the shaft.
2. Shaft fractures generally heal easily and rarely require internal fixation.
3. Malunion is usually not significant since the shoulder is a ball and socket joint, which tolerates some degree of angular, as well as rotational, malalignment.

VI. PELVIC TRAUMA

> *Key Concepts:* In pelvic trauma, all parameters outlined herein must be evaluated since pelvic fractures and dislocations may be extremely subtle. Judet views are very useful, and, for unstable injuries, CT is invaluable. Sacral fractures and/or sacroiliac (SI) joint dislocations are particularly hard to see. Single breaks in the pelvic ring or transverse process fracture at L_5 should alert one to the possibility of a sacral fracture or dislocation.

A. Anatomy: The pelvis is a ringlike structure formed by two arches.
   1. Major arch is posterior and superior, formed by the iliac wings and sacrum, joined at the SI joints.
   2. Smaller arch is anterior and inferior, formed by the pubic and ischial bones, joined at the pubic symphysis.
   3. Three centers of ossification join at the triradiate or Y cartilage of the acetabulum; they may be mistaken for fractures in skeletally immature patients.
   4. *AP*: In a trauma patient it must be demonstrated that the following structures are intact (Fig 3–17):
      a. Iliopubic line.

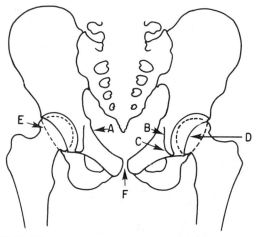

FIG 3–17.
AP pelvis. (Key: A, iliopubic line; B, ilioischial line; C, teardrop; D, anterior acetabular rim; E, posterior acetabular rim; F, symphysis pubis)

        b. Ilioischial line.
        c. Teardrop.
        d. Anterior acetabular rim.
        e. Posterior acetabular rim.
        f. Symphysis pubis.
        g. The sacrum, often obscured by bowel gas, must be observed carefully. The sacral foraminal lines should be checked for distortion, interruption, and asymmetry. A fractured L_5 transverse process particularly suggests an occult sacral fracture. The SI joints are normally wide in adolescents but should be only 2 to 4 mm wide in adults. An increase suggests disruption.

   h. The symphysis pubis width may be up to
      10 mm in adolescents but is no more than
      5 mm in adults. Increased width indicates
      disruption, as does superior displacement.
      (Superior displacement of up to 2 mm is
      normal if the inferior pubic rami remain
      symmetrically placed.)
5. Judet: The lateral pelvis is divided into
   anterior and posterior columns. These
   columns are not easily differentiated on the
   AP film but are important to distinguish for
   reasons of treatment and prognosis. Lateral
   films are impractical because of overlap of
   structures; 45-degree oblique, or Judet, views
   are substituted (Fig 3–18), which show the
   critical anatomy exceptionally well. Note that
   the posterior (or external) oblique shows the
   posterior column and *anterior* acetabular rim,
   while the anterior (or internal) oblique shows
   the *anterior* column and the *posterior*
   acetabular rim. The SI joints and iliac wings
   may also be seen to better advantage. Also
   check the following:
   a. Femoral head subluxation or dislocation
      (usually posteriorly).
   b. Disruption in dome (superior, weight-
      bearing portion) of acetabulum.
   c. Medial migration of femoral head
      (indicating a medial wall acetabular
      fracture).
6. CT:
   a. Should not be used routinely in stable
      pelvic fractures because of increased
      radiation but is particularly useful in
      evaluation for sacral, SI joint fracture-
      dislocations.
   b. It is also useful in a search for intra-
      articular loose bodies after acetabular
      fracture or hip dislocation.
   c. CT has been shown to upgrade the

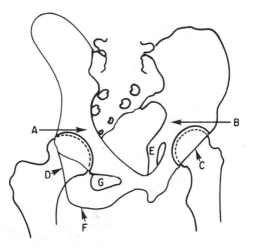

FIG 3–18.
Judet (oblique) view of the pelvis. Note that the *anterior oblique* (right side) shows the *anterior column and posterior acetabular rim* best, while the *posterior oblique* (left side) shows the *posterior column and anterior acetabular rim* best. (Key: **A,** anterior column and iliopubic line. **B,** posterior column and ilioischial line. **C,** anterior acetabular rim. **D,** posterior acetabular rim. **E,** ischial spine. **F,** ischial tuberosity. **G,** obturator foramen.

designation of trauma severity in 30% of unstable pelvic fractures[9] over AP film findings. This usually involved demonstration of severe comminution or displacement at the SI joint, demonstration of occult SI joint or sacral injuries, or demonstration of extension of a fracture into the acetabulum.
   d. In sum, the degree of posterior ring

disruption is often underestimated on plain film alone, and CT may be extremely useful.

B. Sacral fractures.
1. Isolated traumatic sacral fractures usually are transverse and result from a direct blow.
2. When they are a part of complex pelvic trauma, sacral fractures are usually vertical, interrupting the sacral foramina and neural arches.
3. Sacral stress fractures due to osteoporosis are vertical and seen as a dense line, generally extending parallel to the SI joint. They are extremely difficult to see, and a large percentage are missed. They are picked up more frequently by bone scan or CT.

C. Pubic fractures.
1. Ischiopubic synchondrosis: Usually fuse by 12 to 13 years; bulbous, irregular, and asymmetric so may be confused with neoplasm or healing fracture.
2. Insufficiency fractures of the superior and inferior pubic rami are common in osteopenic patients. In the presence of osteoporosis, the pubic fractures may become expanded and appear aggressive or neoplastic. In osteomalacia, the pubic rami often develop wide, lucent Looser's zones.
3. Stress fracture: Occurs at the junction of the pubis and ischium, often during a marathon run. Females more likely to be affected than males.

D. Apophyseal avulsion injuries.
1. Four apophyses: Appear by puberty, fuse by age 25.
   a. Crest of ilium, ending in anterior superior iliac spine (origin of sartorius muscle).
   b. Anterior inferior iliac spine (origin of rectus femoris muscle).
   c. Ischial tuberosity (origin of hamstring muscle).

d. Inferior pubic ramus (at symphysis, origin of adductor muscle).
2. Generally, avulsion results in amorphous bone formation between the pelvis and avulsed fragment, which gradually matures. In the early stages, this ossification may give the appearance of an osteogenic sarcoma.
3. Adductor avulsion injuries of the pubis generally cause sclerosis and widening of one side of the symphysis but can also be seen bilaterally. This may simulate neoplasm or infection.
E. Stable pelvic fractures.
1. Two thirds of pelvic fractures.
2. Single breaks or breaks along the peripheral margins (i.e., avulsion, iliac wing fracture, sacral fracture, or ischiopubic rami).
F. Unstable pelvic fracture-dislocation.
1. One third of pelvic trauma.
2. Involves pelvic disruption in two places or more (double vertical fracture-dislocation).
   a. Most common is vertical shear or Malgaigne; this usually consists of a sacral fracture plus ipsilateral superior and inferior pubic rami fractures. Variants include iliac wing fracture, SI joint disruption, and symphysis pubis disruption.
   b. Another form, the straddle fracture, involves both superior and inferior pubic rami on both sides.
3. These unstable fractures are associated with a significant risk of visceral injury and hemorrhage, and require internal fixation.
4. With unstable pelvic fracture-dislocation, it is common to get an "open-book" disruption of the posterior ring (i.e., loss of the normal angulation of the sacrum with the iliac wings and resultant anterior ring diastasis).

5. On the AP view, the iliac wings may appear asymmetric, one being broader and flatter than the other. The true extent of the "open-book" disruption is best seen by CT (Fig 3–19). The degree of this open book displacement, as well as the amount of cephalocaudad displacement, are indicators of stability and prognosis. If these fractures are not reconstructed internally, change in these parameters must be sought in follow-up films.

VII. HIP TRAUMA

Key Concepts: Subcapital fractures in the elderly may be extremely subtle but are common and must be suspected. Hip dislocations in the presence of other trauma are easily overlooked and, if unreduced for 24 hours, are complicated by AVN. Hip pain in an adolescent should prompt a search for the subtle changes of slipped capital femoral epiphysis.

A. Anatomy.
   1. AP film must show the hip in slight internal rotation. This allows the neck of the femur to be elongated, the greater trochanter to be seen in profile, and the lesser trochanter (a posteromedial structure) to be less prominent.
   2. Frog-leg lateral film superimposes the superior border of the greater trochanter over the femoral neck, simulating an impacted fracture line. It is, however, a useful view to confirm a subcapital fracture, AVN, or slipped capital femoral epiphysis.
   3. Groin lateral is true lateral (Fig 3–20).
   4. Oblique view of the pelvis (Judet) is useful to evaluate for acetabular fracture or a dislocated hip. The internal oblique shows the posterior

AP VIEW

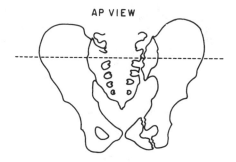

AXIAL VIEW (CT)

NORMAL   OPEN BOOK

FIG 3–19.

"Open book" disruption in a vertical shear fracture. The degree of angular asymmetry of the iliac wings may be difficult to assess on AP film (*top*). The difference between the axial view on normal CT (*left*) and CT with fracture and "open book" displacement (*right*) is much more obvious.

acetabular rim, and the external oblique shows the anterior acetabular rim (see Fig 3–18).
5. Soft tissues: Bulging fat planes may represent a hip effusion, but they are not always present or symmetric. They also require a perfectly positioned patient (external rotation

or flexion of a hip causes a false-positive
bulging of the fat planes). Three fat planes
are iliopsoas, gluteal, and obturator internus.
More reliable indicators of effusion may be an
increased distance from the teardrop,
medially, or the superior acetabulum; in the
absence of trauma, a symptomatic hip with
evidence of effusion must be emergently
aspirated to rule out infection.

B. Epidemiology.
  1. Hip fractures are rare in young and middle-
     aged patients but extremely common in the
     elderly owing to senile osteoporosis[10]: By age
     80, 10% of Caucasian females and 5% of
     Caucasian males fracture a hip; by age 90,
     20% of Caucasian females and 10% of
     Caucasian males fracture a hip.
  2. Subcapital fractures are twice as common as
     intertrochanteric fractures.

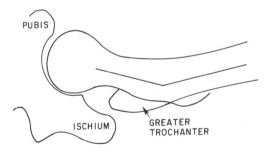

FIG 3–20.
Groin lateral view of the hip, demonstrating the normal
anatomy and the normal neck-shaft anteversion.

3. Femoral neck fractures resulting from falls are highly associated with distal radius and proximal humeral fracture in the elderly.

C. Femoral neck fractures.
   1. Basicervical: Rare; nonunion is a relatively common complication.
   2. Transcervical: Rare.
   3. Subcapital:
      a. Common.
      b. May be impacted or displaced, complete or incomplete.
      c. May be as obvious as a change in angulation or disrupted or angulated trabeculae or as subtle as a line of increased density (impaction) or an irregularity confined to the lateral cortex at the junction of the head and neck.
      d. Ring osteophytes on the AP film on the superior border of the greater trochanter on a frog-leg film may be mistaken for a sclerotic impaction fracture line.
      e. If incomplete or impacted in a reliable patient, may be treated conservatively with bed rest. Otherwise, Knowles pinning is the most common procedure. Depending on the patient's condition and the appearance of the fracture, primary endoprosthesis placement may be elected.
      f. AVN is a complication in 8% to 30% of subcapital fractures.[11]

D. Intertrochanteric fractures.
   1. Common fracture, generally in older age group than subcapital fractures.
   2. Two-, three-, or four-part, depending on involvement of greater and lesser trochanters.
      a. The area of greatest comminution is

posteriomedial, in the regions of the calcar and lesser trochanter. The large defect in this region is seen best on a groin lateral film.

b. The oblique fracture usually angles from the greater trochanter superiorly to the lesser trochanter inferiorly. The reverse diagonal is rare but more unstable.

3. Fixation.
   a. Usually internally fixed with a sliding or dynamic screw. This allows settling and loss of stability with impaction up to a point. If settling continues before the fracture heals, the screw head cuts out of the osteoporotic femoral head and neck.
   b. The optimal placement of the head of the screw is slightly posterior and inferior in the femoral head, with the tip approximately 6 mm from the articular surface. The screw head specifically should *not* be anterior or superior.
   c. Three common positions are accepted for reduction (Fig 3-21).
      (1) Anatomic: If the posteromedial comminution is not severe.
      (2) Medial impaction: The shaft is medially displaced and impacted on the spike of the proximal fragment; this gives stability at the expense of a minor loss of length.
      (3) Valgus: Osteotomy to reduce shear and thereby stabilize the fracture.

4. AVN is rare. The major complications are instability and cutting out of the hardware as the pattern of fixation collapses into a varus configuration.

ANATOMIC

A

MEDIAL
DISPLACEMENT
WITH IMPACTION

B

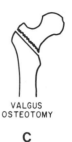
VALGUS
OSTEOTOMY

C

FIG 3–21.
Positions accepted for reduction of an intertrochanteric
fracture: **A,** anatomic; **B,** medial displacement (of the
shaft) with impaction (of the medial spike on the proximal fragment). **C,** valgus osteotomy.

 E. Avulsion fracture of the lesser trochanter.
  1. Not uncommon in children or adolescents, as
   avulsion of the apophysis.
  2. If it is found as an isolated fracture in adults,
   it is usually due to underlying bone
   pathology (metastatic disease).
 F. Femoral shaft fracture.
  1. May be fixed internally by compression plate,
   with or without cerclage wire if highly
   comminuted.
  2. Intramedullary (IM) rod may be used if the
   fracture is not highly comminuted and does
   not have a large butterfly fragment.
   Successful use of an IM rod depends on its
   filling the width of the medullary canal to
   stabilize a fracture. If the fracture is at the
   proximal or distal third (where the canal
   flares), rotational stability may be lost unless
   the rod is fixed with an interlocking screw.

   3. In evaluating an IM rod, check for distraction
      of the major fracture fragments, progressive
      shortening (with migration of the rod), and
      rotation of the fragments.
G. Stress fracture of the femur.
   1. In the proximal or midshaft, it usually
      involves the medial cortex.
   2. In the distal third of the shaft, it usually
      involves the posterior cortex.
H. Hip dislocation.
   1. A rare result of severe trauma.
   2. Often associated with femoral shaft fracture.
      The dislocation may be overlooked clinically
      because of the discomfort of the obvious shaft
      fracture. It may be overlooked
      radiographically as well unless a pelvic film is
      obtained.
   3. Posterior dislocation is most common (~90%).
      The head is usually located superiorly and is
      held in internal rotation. There is usually an
      associated posterior acetabular rim fracture. If
      the dislocation is directly posterior, it may be
      more difficult to recognize on the AP film: a
      lack of congruence of the head with the
      acetabulum may make the diagnosis.
      Alternatively, the femoral heads may appear
      to be different sizes: the smaller one is less
      magnified, therefore closer to the film, and
      therefore posterior.
   4. In the rare anterior hip dislocation, the head
      is usually found overlying the obturator
      foramen.
   5. Early diagnosis is imperative since a delay in
      reduction sharply increases the probability of
      avascular necrosis (approaching 50% if
      unreduced for 24 hours).
   6. After reduction, the hip joint should be

studied closely for an increase in teardrop
width. Such an increase may indicate the
presence of retained fracture fragments,
which is easily confirmed with CT.
I. Slipped capital femoral epiphysis (SCFE).
  1. Epidemiology.
      a. Usually 10 to 16 years old. Rarely it may
         be seen in younger children in the
         presence of infection, severe trauma,
         congenital hip dislocation, or rickets.
      b. Males are more often affected than
         females.
      c. Blacks are more often affected than
         Caucasians.
      d. Obese persons are more often affected
         than nonobese persons.
      e. Bilateral 20% but rarely symmetric.
  2. SCFE occurs during the years of rapid
     growth, which is also the stage at which the
     femoral neck configuration changes from
     valgus to varus. This introduces the factor of
     shear stress in a growth plate weakened by
     rapid growth; minor trauma may therefore
     precipitate SCFE.
  3. Radiographic appearance on AP film (Fig 3–
     22) (SCFE almost always is posteromedial):
      a. The epiphyseal plate appears wider, with
         less distinct margins.
      b. The epiphysis itself appears shorter.
      c. A line drawn along the lateral femoral
         neck may intersect a smaller portion (or
         none) of the femoral head.
      d. A frog-leg lateral or groin lateral confirms
         the findings.
  4. Treated by pinning in situ; this yields a varus
     deformity with a short, broad femoral neck.
  5. Complications.

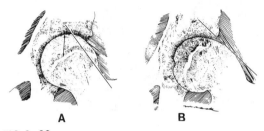

FIG 3–22.
AP of a normal hip (**A**) and slipped capital femoral
epiphysis (**B**) showing the medial displacement and
"shortening" of the epiphysis, and "widening" of the
epiphyseal plate. A line drawn along the lateral aspect
of the femoral neck usually intersects a portion of femoral head (**A**) but often does not intersect the femoral
head in a slipped capital femoral epiphysis (**B**).

    a. DJD: Surprisingly, a late occurrence, often
       30 years later.
    b. AVN: In about 10%; the probability is
       greater with open reduction, acute severe
       slips, and attempted repositioning.
    c. Chondrolysis: Acute disappearance of
       cartilage in SCFE, chondrolysis has been
       associated with pin penetration through
       the articular cortex. In one study,
       chondrolysis was seen more often in black
       girls with SCFE; chondrolysis is also seen
       idiopathically. The differential diagnosis is
       infection.

VIII. KNEE TRAUMA

> **Key Concepts:** Most knee injuries involve soft tissues and show only effusion on plain film. A fat-blood level in suprapatellar pouch demands a careful search for intracapsular fracture. Meniscal anatomy is somewhat complex; this must be borne in mind when interpreting knee arthrograms or MR studies. Knee dislocations have commonly associated arterial injury.

A. Soft tissues in knee trauma.
   1. On lateral film, the fat pad posterior to the quadriceps tendon is divided into anterior and posterior compartments by a soft tissue density, the suprapatellar bursa.
      a. In the absence of effusion, the suprapatellar bursa is less than 5 mm wide.
      b. Suprapatellar lipohemarthrosis indicates an intracapsular fracture, which may be occult.
B. Femoral condyle.
   1. If intra-articular femoral condylar fracture is present, patient is at risk for avascular necrosis.
   2. Osteochondritis dissecans.
      a. Etiology uncertain but possibly due to repeated minor trauma.
      b. Most often involves lateral portion of medial femoral condyle.
      c. Most common in adolescents and young adults.
      d. Arthrography demonstrates whether the overlying cartilage is intact.
      e. Normal variants in children: In 3- to 6-year-olds, the femoral epiphysis is normally irregular, especially medially. In 10- to 13-year-olds, a femoral condylar irregularity is found on both condyles posteriorly (see best on a notch view); it is

bilateral, asymptomatic, and resolves with maturity. Neither of these entities should be mistaken for osteochondritis dissecans or an erosive arthropathy.

3. Spontaneous osteonecrosis: A entity distinct from osteochondritis dissecans; affects an older adult age group. Pain is a much more striking feature; found on the medial femoral condyle, but more medial and superior than osteochondritis dissecans; subchondral lucency with subsequent flattening; 75% have associated torn medial meniscus.

C. Tibial fracture.
   1. Tibial plateau fracture.
      a. Often seen in auto-pedestrian accidents since the plateau is at the height of fenders and bumpers.
      b. Eighty percent are limited to the lateral plateau since most result from a valgus stress.
      c. The tibial plateaus are sloped posteriorly 10 to 20 degrees. The AP knee film therefore is not tangential to the tibial joint surface.
      d. Anterior depressed fragments may be overlooked since the anterior margin is projected superior to the posterior margin. Similarly, the extent of posterior depressed fragments may be exaggerated.
      e. Oblique films may be necessary for the initial diagnosis. Tomography may be required to establish the extent of tibial plateau depression.
      f. In general, depression greater than 1cm or widely separated (5 mm) vertical split fractures require internal fixation.
   2. Anterior tibial tubercle apophysis: May have multiple ossification centers and appear fragmented. The osteonecrosis Osgood-Schlatters disease should be suggested only

in the presence of fragmentation, soft tissue
swelling, and pain.

3. Anterior tibial spine fracture.
   a. Adjacent to the site of origin of anterior
      cruciate ligament, so may be avulsed.
   b. More common in children and adolescents
      than adults.
   c. May appear hinged, completely detached,
      or inverted.

4. Avulsion of the posterior cruciate ligament
   may result in avulsion of a sliver of bone
   from the posterior tibia near the plateau.

5. Lateral capsular sign: Avulsed sliver of bone
   from the tibia, at site of capsular insertion.
   Sign associated with ACL tear.

6. Proximal fibular fracture or dislocation.
   a. Dislocation is rare.
   b. Normally, the fibular head is located
      posterolateral to the tibia, so there is slight
      overlap of the two on both AP and lateral
      films.
   c. Injury may involve peroneal nerve.
   d. Site of insertion of the lateral collateral
      ligament, so may be avulsed.

7. Stress fracture.
   a. Common in the proximal tibial shaft.
   b. Early, may see faint transverse or oblique
      lucency within the posterior cortex of the
      proximal tibial shaft.
   c. Later, see transverse band of density and
      callus formation along the posterior (not
      anterior) cortex.
   d. Radiographic lag of 2 to 6 weeks.

D. Epiphyseal injury.
   1. Occurrence is relatively rare about the knee,
      but complications are frequent.
   2. Salter 2: 70%.
   3. Salter 3: 15%: These usually involve the
      medial condyle and are due to valgus stress.
      They are undisplaced and often occult.
      Oblique or valgus stress films are helpful.

4. The knee is the most common site of Salter 5 fractures: They usually are seen in the proximal tibia, associated with tibial shaft fractures.

5. A disproportionately large number of significant growth disturbances stem from epiphyseal injuries about the knee, despite the rarity of epiphyseal injury. Therefore the prognosis should be guarded, and early diagnosis of bony bridging across the epiphysis should be actively sought.

E. Patellar trauma.

1. Patellar fracture.

   a. Sixty percent are transverse, through the midportion. These are due to an indirect force (violent pull of the quadriceps tendon).

   b. Twenty-five percent are stellate, due to direct trauma.

   c. Vertical much less common.

   d. Bipartite or multipartite patella: The fragments are found on the superolateral border and have well-corticated margins. They are frequently, but not invariably, bilateral.

   e. Dorsal defect of the patella: A rounded lucency on the articular (dorsal) side of the patella is a normal variant, not to be confused with osteochondritis dissecans.

   f. Osteochondral (flake) fracture: Usually from the medial facet, associated with lateral patellar dislocation and seen on the sunrise view.

2. Patellar dislocation: Usually lateral; tendency is defined by patellar tilt, lateral patellar displacement (Fig 3–23), and patella alta.

   a. Study of choice is the *sunrise* view: Obtained with the knees flexed 20 degrees. (With more flexion, the patella becomes engaged in the patellofemoral

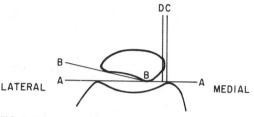

FIG 3–23.
Sunrise view, obtained to evaluate patellar tilt or displacement. A—A and B—B form an angle that normally is open laterally. If the angle is open medially, this constitutes patellar tilt. Line C is perpendicular to A—A at the tip of the medial femoral condyle. Line D is 1 mm lateral to line C (line D normally intersects the medial patella). If the patella is lateral to line D, it is laterally displaced and subject to subluxation; this position occurs in 30% of patients with chondromalacia patellae. [12]

     groove, and subtle abnormalities in alignment are not detected.)

  b. In most cases, the lateral facet is slightly longer than the medial facet.

  c. Patella alta: Elongation of the infrapatellar tendon associated with recurrent subluxation. On the lateral flexed (20 to 30 degrees) view of the knee, the ratio of the infrapatellar tendon length (from the inferior pole of the patella to the anterior tibial tubercle) to the length of the patella is 1.0 ± 0.2. If this ratio is greater than 1.2, patella alta exists.

F. Knee dislocation.

  1. Anterior dislocation is more common than posterior or medial-lateral.

  2. Thirty percent have an associated arterial

injury (since the popliteal artery is fixed both proximal and distal to the knee joint), so arteriography is mandatory.

G. Knee arthrograms.

1. Arthroscopy and MRI have replaced most of these studies. Current indications are a questionably abnormal knee that may not require arthroscopy; a clinically abnormal knee with a normal arthroscopic examination; tears of partially resected meniscal remnants; diagnosis of Baker's cyst.

2. Tears are most commonly seen in the posterior horn and body of the medial meniscus, so these should be examined first.

3. Discoid meniscus: Usually involving the lateral meniscus that is much wider than normal. The normal crescent shape is absent, and tears, with locking, are common. It is a likely diagnosis in children under 10 years of age who have signs of an internal knee derangement.

4. O'Donoghue's terrible triad: Anterior cruciate ligament tear, medial collateral ligament tear, medial meniscus tear.

5. MRI now can replace most knee arthrograms if done carefully with a technique capable of thin sections and excellent resolution. Radial GRASS imaging gives an appearance similar to arthrography.

6. Some anatomic points are worth remembering when interpreting knee arthrographs or knee MR images:

   a. The medial meniscus has a very long triangular posterior horn which gradually becomes smaller, so that it is shaped like an equilateral triangle in the body. There is a normal superior recess at the posterior horn.

   b. The lateral meniscus is also large posteriorly, becoming more equilateral at the body. There is a normal inferior recess anteriorly. The popliteal hiatus may be seen over a relatively large area, from the posterior body through most of the posterior horn.

IX. ANKLE TRAUMA

> *Key Concepts:* Occasionally lateral malleolar fractures are seen only on the lateral film. Impaction at the corners of the plafond is an easily missed injury. Anatomic reduction (including normal length of fibula) is crucial for an acceptable result. With an isolated medial injury and widened mortise, proximal fibular fracture should be sought. Two special cases are found in the maturing skeleton—juvenile Tillaux[1] and triplane fractures.

A. Anatomy.
   1. The medial malleolus has two colliculi. The anterior one is longer than the posterior one, giving it a double density on an AP film.
   2. The lateral margin of the tibia has the fibular notch. The fibula normally fits the notch snugly. If, after a fracture, the fibula is shortened, the more bulbous distal fibula does not articulate properly in the fibular notch. This leads to lateral shift of the talus, decreased surface contact between the tibia and talus, and early degenerative joint disease (DJD; see Fig 3–24); therefore, only 2 to 3 mm of fibular shortening is acceptable.
   3. The lateral malleolus is located 1 cm distal and 1 cm posterior to the position of the medial malleolus.
   4. The tibial joint surface is termed the plafond (ceiling).
   5. A 15- to 20-degree internal oblique film brings

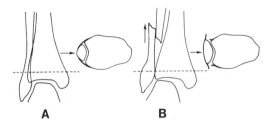

FIG 3–24.
AP view of the normal ankle, **(A)** with axial cut through the fibular notch. After fracture and fibular shortening, the larger distal fibula does not articulate as well with the fibular notch and ankle mortise widening occurs, **(B)**.

the malleoli parallel to the horizontal plane for evaluation of the ankle mortise. Joint space should be an even 3 to 4 mm over the entire talar surface. Two millimeters' widening of the mortise is abnormal.

6. The articular surface of the talus narrows posteriorly. This prevents posterior dislocation, which requires disruption of the ankle mortise.

7. The base of the fifth metatarsal is included on the lateral ankle film since it is a common site of a fracture that may be mistaken clinically for ankle trauma.

8. For stress films, varus and valgus stress is applied to the calcaneus. An anterior drawer stress may also be performed. There is a wide range of normal ankle laxity (normal talar tilt with stress is 10 to 12 degrees but may increase to 20 degrees in lax ankles), so comparison ankle stress films are recommended.

9. Normal variants include the os trigonum

posterior to the talus (fused or unfused) and irregular and multiple accessory ossicles at both medial and lateral malleoli.

10. Ankle effusion is seen on lateral film as an anterior convex soft tissue density at the tibiotalar joint. The pre-Achilles fat triangle lies between the Achilles tendon and the deep muscles of the leg and is sharply defined in the absence of effusion or inflammation. The fat density extends 2 mm distal to the posterior tubercle of the calcaneus.

B. Trauma:

1. Structures to be examined:
   a. Soft tissues, for swelling or effusion; soft tissue swelling distal to the malleoli suggests ligamentous injury.
   b. Malleoli and posterior tibia, for fractures.
   c. Mortise, for disruption.

2. In general, oblique or spiral fractures of the malleoli are due to impacting (pushing) forces, while transverse fractures are due to avulsing (pulling) forces. However, rotational forces are usually a complicating factor to simple inversion or eversion injuries.

3. Lauge-Hansen classification is based on the mechanism of occurrence and may be useful to focus the search for less obvious injuries. It is not useful for treatment or prognosis.

4. Weber (AO) classification is the one most used by surgeons since it correlates well with both treatment and prognosis. It uses the level of the fibular fracture to deduce the injury to the tibiofibular ligaments.
   a. Weber A: Transverse avulsion fracture of the lateral malleolus at or distal to the tibiofibular joint. This injury spares the tibiofibular ligament complex. It may be associated with an oblique fracture of the medial malleolus.

    b. Weber B: Spiral fracture of the lateral malleolus beginning at the level of the ankle joint. This leads to a partial disruption of the tibiofibular ligament, and diastasis of the ankle mortise depends on the extent of injury. It may be associated with a transverse or slightly oblique avulsion fracture of the medial malleolus below the ankle joint or with a deltoid ligament rupture.

    c. Weber C: Fibular fracture proximal to the ankle joint. This invariably tears the tibiofibular ligament complex and leads to lateral talar instability. It may be either a pure ligamentous tear or an avulsion of the anterior (Tillaux-Chaput) or posterior (Volkmann) tubercles of the distal tibia or, more rarely, a flake from their fibular attachment. The medial malleolus is avulsed just below the level of the ankle joint, or the deltoid ligament may be torn.

C. Postoperative evaluation: Exact anatomic and stable reconstruction of the ankle mortise is necessary to prevent traumatic arthritis:

    1. Correct length of fibula and its exact position in the fibular notch of the tibia.

    2. Restoration of the ankle mortise.

    3. Less than 1 to 2 mm displacement of either a posterior malleolar fragment or posterior displacement of the distal fibular fragment. When present, these are signs of persistent subluxation of the ankle joint.

D. Other fractures of ankle.

    1. Watch for *impaction of the plafond* at its junction with the medial malleolus in an inversion injury. This may indicate a severe injury and may be quite occult, as may be osteochondral fractures of either the medial or lateral dome of the talus.

    2. With an eversion injury and isolated

transverse medial malleolar fracture or isolated widening of the ankle mortise there may be a tibiofibular ligament tear and fracture of the proximal third of the fibula (*Maisonneuve* fracture). The entire fibula should be examined radiographically in such a circumstance since the pain of the proximal fibular fracture may be masked by the pain of the ankle injury.

3. *Insufficiency* fractures may be seen simultaneously in the distal tibia and fibula, usually within 3 to 4 cm of the plafond.

4. *Stress* fractures are seen in the fibulas of runners 3 to 7 cm from the tip of the lateral malleolus.

5. Fusion of the distal tibial epiphysis starts at 12 to 13 years of age. It begins centrally and proceeds medially and finally laterally. This pattern may be confused on radiographs with a fracture. The *juvenile Tillaux* fracture is seen secondary to this fusion pattern; it is a Salter 3 fracture of the lateral portion of the distal tibial epiphysis that occurs after fusion of the medial portion of the epiphysis (Fig 3–25).

6. The *triplane* fracture is another juvenile pattern of ankle fracture. It involves the lateral half of the distal tibial epiphysis and a posterior triangular metaphyseal fragment. "Triplane" indicates the three planes of the fracture: vertical through the epiphysis, horizontal through the epiphyseal plate, and oblique through the metaphysis (extending from anterior and inferior at the epiphyseal plate, posteriorly and superiorly; Fig 3–26). There are two types. If a triplane fracture occurs after the medial portion of the epiphysis has fused, the medial malleolus remains intact and it is a two-fragment triplane fracture. If the triplane fracture occurs before the epiphysis starts to fuse, a

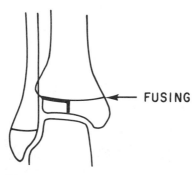

FIG 3–25.
Juvenile Tillaux' fracture: Salter 3 injury occurring at the last portion of the epiphysis to fuse.

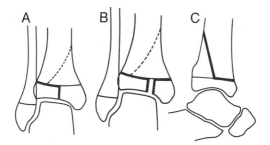

FIG 3–26.
Triplane fractures. **A,** two-part triplane, with fusion of the medial portion of the epiphysis. **B,** three-part triplane occurring prior to fusion of the medial portion of the epiphysis. **C,** lateral appearance of either variety of triplane fracture, which in this view looks like only a Salter 2.

three-fragment fracture occurs; it gives the appearance of a Salter 2 involving the posterior metaphysis and a Salter 3 involving the medial epiphysis. With reconstruction, CT is very useful in completely delineating the extent of fractures and displacement.

X. FOOT TRAUMA

> ***Key Concepts:*** Calcaneal fracture is intra-articular if Boehler's angle is decreased. Calcaneal fractures are associated with lumbar spine fractures and are often bilateral. Tarsometatarsal (TMT) fracture-dislocations are very difficult to diagnose and require attention to precise anatomy at those joints. A child's stubbed toe may develop into a Salter 2 fracture and osteomyelitis.

A. Anatomy.
   1. The calcaneus is tent-shaped on lateral. This is described by Boehler's angle, normally 28 to 48 degrees.
   2. The calcaneal apophysis and immature tarsal navicular normally appear dense and often fragmented. This is a normal variant rather than fracture or AVN.
   3. The articulations of the tarsals and metatarsals are very precise and must be observed carefully to rule out midfoot (Lisfranc) fracture-dislocation (Fig 3–27):
      a. AP films: Lateral border of the first metatarsal aligns with the lateral border of the medial cuneiform; medial border of the second metatarsal aligns with the medial border of the middle cuneiform.
      b. Oblique films: Lateral border of the third metatarsal aligns with the lateral border of the lateral cuneiform; the medial border of the fourth metatarsal aligns with the

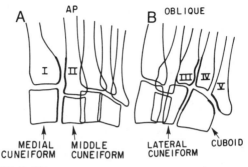

FIG 3–27.
Normal alignment of the tarsometatarsal joints, as outlined in the text. The first and second are evaluated on the AP film, while the third, fourth, and fifth are evaluated on the oblique film. The bold lines indicate the surfaces of the tarsometatarsals which must align with one another on each view. The alignment must be precise, and the lateral film often appears normal even with severe derangement of these joints.

        medial border of the cuboid; the base of the fifth metatarsal articulates with the cuboid but normally extends lateral to its lateral border.
  c. The base of the second metatarsal is recessed between the medial and lateral cuneiform in a lock-and-key configuration.
4. The apophysis of the base of the fifth metatarsal is longitudinally oriented and must not be mistaken for Jones' fracture. It may also be normally bipartite.
5. The epiphysis of the proximal phalanx of the great toe may be bifid, simulating fracture.
6. Normal irregularities and excrescences may be seen along the shafts of all the phalanges.

7. Sesamoid bones may be bipartite or multipartite, simulating fracture.
8. Accessory ossicles are common and are not invariably bilateral. They are well-rounded and corticated, which helps differentiate them from fractures. The most common in the foot are the os peroneum (adjacent to the cuboid), os trigonum (on the lateral film, posterior to the talus), and os tibiale externum (on the AP film, adjacent to the navicular). Others are diagrammed in Keats' *Atlas of Normal Roentgen Variants That May Simulate Disease.*
9. The talocalcaneal joint has three different facets and appears least complicated when imaged by CT. It is completely described in Chapter 5 (Tarsal Coalition).

B. Trauma.
1. Calcaneus.
   a. Fractures sometimes are seen best on axial (Harris) view.
   b. Classified as intra- or extra-articular; a decreased Boehler's angle implies an intra-articular calcaneal fracture.
   c. Bilateral 10%.
   d. Ten percent are associated with thoracolumbar fractures (Don Juan fractures) since common mechanism is a fall from a height.
   e. Stress fracture may be seen only as a vertical linear density 10 to 14 days after onset of symptoms.
2. Talus.
   a. Other than chips and avulsions, the most common fracture is vertical neck; may be associated with talar dislocation.
   b. Very susceptible to AVN.
   c. Osteochondral fractures or osteochondritis dissecans may be seen on dome of talus either medially or laterally. They are most often seen in lax ankles, and

osteochondritis dissecans is not uncommonly bilateral.

3. Navicular: Stress fractures are rare but tend to occur in joggers, are horizontal, and usually require tomography to diagnose.

4. Lisfranc fracture-dislocation.
   a. Dorsal dislocation of tarsometatarsal joints, usually associated with several chip and avulsion fractures.
   b. Precise study of these anatomic relationships on the oblique and AP films may be required for diagnosis. Lateral film may appear normal.
   c. Classified as homolateral (usually lateral dislocation of metatarsals 2–5 or 1–5) or divergent (lateral dislocation of metatarsals 2–5 and medial subluxation of the first metatarsal). Early radiographic findings may be extremely subtle, with only a slight widening and offset at the first and second metatarsals; with continued weight-bearing, further separation occurs.
   d. Lisfranc fracture dislocations are much more commonly a manifestation of a diabetic Charcot (neuropathic) joint than due to trauma; lateral film may appear normal.

5. Jones' fracture: Transverse fracture at base of fifth metatarsal (also called dancer's fracture); remember that the apophysis is oriented longitudinally and should not be mistaken for a fracture.

6. Metatarsal stress fracture (march fracture).
   a. Most common stress fracture.
   b. Second or third metatarsal most common.
   c. Nondisplaced and often not radiographically apparent until 7 to 10 days after injury, when fluffy callus formation or periosteal reaction is seen.

7. Stubbed great toe in a child.

   a. Nail bed of great toe is attached to
      periosteum at the level of the proximal
      metaphysis.
   b. Stubbed great toe with a nail bed injury
      not infrequently results in a Salter 1 or 2
      fracture and/or osteomyelitis.

XI. SPINE TRAUMA

> *Key Concepts:* Cervical injuries may be ex-
> tremely subtle, yet clinically devastating. Abnor-
> mal prevertebral soft tissues are usually an
> extremely reliable sign of significant injury. Nu-
> merous normal variants may suggest fracture.
> Each examination must be individualized.

Cervical Spine

A. Normal anatomy relevant to trauma.
   1. Lateral film: Initial film, to be cleared before
      further films are taken: *Must* see to top of T_1:
      a. Prevertebral soft tissues:
         (1) In adults:
             a Not more than 5 mm at C_3 and C_4.
             b Less than 22 mm at C_6.
         (2) In children:
             a Two-thirds the width of C_2 body at
                C_3 and C_4.
             b Not more than 14 mm at C_6.
      b. Cervical lordosis: Loss of the lordosis may
         represent muscle spasm; however, it is
         normally absent in 20% of patients in the
         neutral position. It is absent in 70% of
         normal patients if the chin is depressed
         only 1 in.
      c. Four continuous curves describe the
         normal position of the bony elements (Fig
         3–28):
         (1) Anterior vertebral body line.
         (2) Posterior vertebral body line; exception

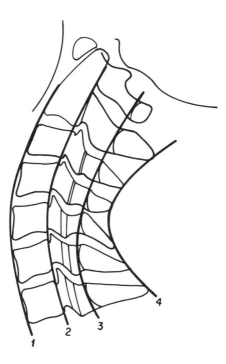

FIG 3–28.
Lateral cervical spine with four normal contiguous curves:
1, anterior vertebral line; 2, posterior vertebral line; 3,
spinal laminar line; 4, posterior spinous line.

to this is in children, where there is
often a physiologic offset of 2 to 3 mm
of C_2 on C_3 or C_3 on C_4.
(3) Spinal laminar line.
(4) Posterior spinous process line.

d. In the absence of disk disease, the distance between adjacent posterior vertebral bodies is uniform at all levels. A gap at one level suggests posterior ligamentous injury; this would be supported by an abnormal fanning of the spinous processes. Note that fanning of the spinous processes normally is not uniform: Normally it is greater for the proximal and distal cervical elements than for the middle elements. Therefore, fanning of C_{3-5} spinous processes is indicative of posterior ligamentous injury (Fig 3–29,A).

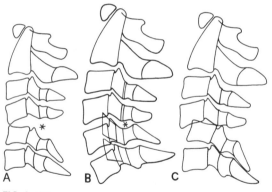

FIG 3–29.
A, lateral cervical spine with fanning of spinous processes and a gap at the posterior vertebral body, indicating posterior ligamentous injury. **B,** abnormal rotation of facets with an abrupt transition at C_{4-5}, indicating a rotational subluxation. **C,** locked facets, with interruption of all four curves and total lack of contact between the articular surfaces of the facets at C_{4-5}.

e. The facets are bilateral. Since lateral films are usually not positioned perfectly, there is often overlap of the right and left facets at each level. This should be uniform at all levels in the absence of rotation. An abrupt change in amount of overlap indicates an abnormal rotation (Fig 3–29,B).
f. The inferior articular surface of the facet should be in full contact with the more distal element's superior facet articular surface. The absence of such full contact indicates a subluxed, perched, or locked facet (Fig 3–29,C). A unilateral locked facet implies abnormal rotation as well.
g. Odontoid process (dens) is normally tilted posteriorly on the body of C_2.
h. Atlantoaxial distance is measured at the base of the dens between the anterior cortex of the dens and the posterior cortex of the atlas' anterior arch. In adults, this distance is not more than 2.5 mm and does not change with flexion. In children, the distance may be as great as 5 mm and may change by 1 to 2 mm with flexion.

2. AP film: May provide a valuable clue to spinous process avulsion. The spinous processes should form a continuous line. In clay-shoveler's fracture, a "double" spinous process is caused by the slightly displaced fragment of the tip overlying the base of its process (see Fig 3–30).
3. Oblique film: Used to evaluate posterior elements for fracture and confirm normal overlap of facets.
4. Open-mouth film.
   a. Used to evaluate the odontoid process for fracture. Do not confuse the bottom of the incisors or the arch of the atlas with a fracture.

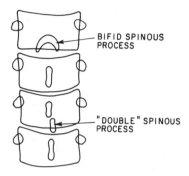

FIG 3–30.
AP cervical spine showing a normal bifid spinous process and an abnormal "double" spinous process, indicating a clay-shoveler's fracture. Note that facet joints are not seen on the AP because of their normal angulation.

      b. Used also to evaluate the integrity of the ring of the atlas. In neutral position, there is exact alignment of C_1 on C_2 without offset of the facets (i.e., equal distance from the dens to the medial margin of each facet). With rotation, the atlas moves as a unit: the lateral facet of C_1 is offset laterally on one side of the dens and medially on the other. Bilateral lateral offset of the facets indicates a C_1 ring fracture in adults; in children, it may be a normal variant.

   5. Other films to be tailored to the individual patient's needs:

      a. Lateral flexion-extension views.

        (1) If a fracture is not seen but there are other suggestions of ligamentous injury (usually abnormal gapping of the

bodies posteriorly or fanning of the spinous processes on the lateral film), flexion-extension films will help evaluate the extent of the injury and the degree of stability.

(2) The patient is allowed to flex and extend alone, without force. A physician should supervise the filming, and the patient should be awake, cooperative, and neurologically intact.

(3) Watch for increase or reduction of the posterior splaying. Also evaluate facet subluxation.

b. Pillar views.

(1) AP with 20 to 30 degrees' caudad angulation.

(2) Used to profile the facets (they are normally angulated with respect to the AP film).

(3) Watch for facet compression or fracture. These are especially at risk in a hyperextension injury.

c. Tomograms/CT.

(1) Either may be useful in an individual situation: tailor to the problem.

(2) Tomograms require excellent technique and patient cooperation.

(3) CT may miss fractures in the axial plane (especially around C_1 and base of dens) if thin section with reconstruction is not utilized and must be interpreted with care since there is overlap of adjacent levels posteriorly.

B. Congenital anomalies and normal variants.

1. Fusion or lack-of-segmentation anomalies are common.

2. Occipitalization of the atlas: Lack of segmentation at atlantooccipital junction; presents with atlantoaxial subluxation,

simulating a traumatic disruption. There is also an abnormally large gap between the spinous processes of C_1 and C_2, with the atlas located unusually close to the occiput. The diagnosis is established by a flexion film, which demonstrates fixation of the atlas to the occiput. In addition, the odontoid often has a bizarre shape.

3. Absence or lack of fusion of ossification centers is especially confusing at C_1 and C_2.

   a. Normal ossification of C_1: The body (occasionally bifid) and each of the neural arches (Fig 3–31); posterior arch defects are common in C_1. Also, the synchondrosis between the body and arch may fuse asymmetrically (usually by age 7 years).

   b. Normal ossification of C_2 (Fig 3–32): Four ossification centers (one for each neural arch, the body [occasionally bifid], and the odontoid process). The body-neural arch synchondroses fuse asymmetrically between age 3 and 6; the body-odontoid

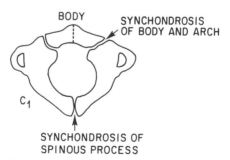

BODY

SYNCHONDROSIS OF BODY AND ARCH

C_1

SYNCHONDROSIS OF SPINOUS PROCESS

FIG 3–31.
Normal ossification centers of C_1, indicating sites at which lack of fusion may simulate a fracture.

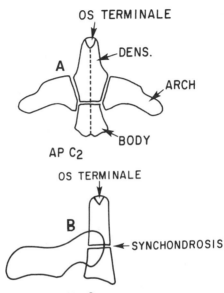

FIG 3-32.
Normal ossification centers of C_2 — AP **(A)** and lateral **(B)**— indicating sites at which lack of fusion may simulate a fracture.

process synchondrosis also fuses between 3 and 6 years. A persistent lucent synchondrosis may remain into adult life, located well *below* the level of the apparent "base" of the odontoid. This is differentiated from an odontoid fracture, which usually occurs at the true base of odontoid.

c. Os terminale: An ossification center at the

superior tip of the odontoid process.
Before it ossifies, the tip of the process is
V-shaped. The os terminale normally fuses
by age 12 (Fig 3–33,A) but may persist
unfused, simulating a tip of odontoid
fracture (Fig 3–33,B).
d. Os odontoideum: In the presence of a
hypoplastic odontoid, the os terminale
becomes overgrown and is termed the os
odontoideum. It is a large ossicle which is
separated from the hypoplastic odontoid

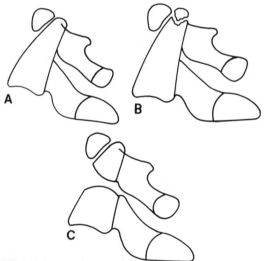

FIG 3–33.
Spectrum of os terminale variants: **A,** normal fusion. **B,**
a simple unfused os terminale. **C,** os odontoideum, the
result of a hypoplastic odontoid and overgrown os
terminale.

by a wide gap and which moves with the arch of the atlas. It may appear quite bizarre (Fig 3–33,C).
4. A bifid spinous process may project into the neural foramen on an oblique film, simulating a fracture.
5. A transverse process may be elongated, appearing as a bone fragment on the lateral film.
6. An enlarged uncinate process may simulate a vertebral body fracture on the lateral film. Confirm on the AP as DJD of the uncovertebral joint.
7. In a child, the ring apophysis (seen especially well anterosuperiorly on the lateral film) may simulate a chip or avulsion fracture. Before the ring apophysis is ossified, it appears as a corner notch in the vertebral body and may simulate anterior body wedging.
C. Injury pattern of the cervical spine.
1. C_1 (atlas):
a. Bilateral vertical fracture through the neural arch is most common; to be differentiated from congenital defects.
b. Jefferson's fracture: Vertical compression produces a burst fracture involving both anterior and posterior arches. It is stable, and usually the patient is neurologically intact unless there is also disruption of the transverse ligament.
c. Atlantoaxial rotary displacement: A rotatory locking of the facets, usually seen in childhood, presenting as torticollis. Most patients recover spontaneously.
(1) Open-mouth view: One lateral mass of C_1 appears wider and closer to the midline. The opposite is narrower and laterally offset. One of the facets may be obscured by overlapping.
(2) Lateral view: Posterior arches of the

atlas fail to superimpose because of head tilt; may have wider atlas-dens space.

d. CT often is useful.

2. C_2 (axis):

   a. Odontoid (dens) fracture: Usually through the base of the dens; if it has an oblique extension into the vertebral body anteriorly, it may not be seen on the open-mouth view; must be differentiated from congenital variants (see B, above).

   b. Hangman's fracture: Hyperextension injury resulting in bilateral neural arch fractures. The odontoid and its attachments are intact; nerve damage is uncommon owing to the width of the canal at this level.

3. Flexion injuries:

   a. Anterior wedge: Relatively minor injury, not usually associated with posterior retropulsion of the body.

   b. Hyperflexion sprain-anterior subluxation: Posterior ligamentous complex disrupted; localized increased height of intervertebral disk space, associated with fanning of spinous processes and a local kyphotic angulation. These findings are accentuated on flexion films; may allow facet subluxation or locking. There is delayed instability in 20%!

   c. Unilateral locked facet: Due to flexion, distraction, and rotation; an abrupt change in amount of facet overlap on lateral film. Most common locations are C_{4-5} and C_{5-6}; 35% are associated with fracture (usually facet).

   d. Bilateral locked facet: Due to flexion with enough distraction for facets to become disarticulated; the vertebral body is displaced approximately 50% of the body length on the lateral film. Both lateral and oblique films show the "jumped," locked facets; high incidence of cord damage.

   e. Clay-shoveler's fracture: Avulsion of the spinous processes (usually C_6 or C_7) due to flexion; "double" process seen on AP.

   f. Teardrop burst fracture: Most severe flexion fracture compatible with life; comminuted vertebral body fracture with triangular fragment from the anteroinferior border of the body. The posterior body is displaced into the spinal canal with high probability of neural damage.

4. Extension injuries (signs may be quite subtle):

   a. Prevertebral soft tissue swelling.

   b. Posterior body displacement.

   c. Widened intervertebral disk space, especially at the anterior portion of the bodies.

   d. Vacuum phenomenon at the anulus fibrosus (in the absence of degenerative disease) is highly suggestive of anterior soft tissue injury. Avulsion fracture from the anteroinferior margin, especially of C_2 or C_3. Facet (pillar) compression fracture may be unilateral and may require pillar views for diagnosis. May result in nerve root compression; not uncommonly, spinal cord injury occurs without fracture or dislocation.

5. Radiographic signs of instability:

   a. Spinous process fanning.

   b. Widening of intervertebral disk space (in neutral, flexion, extension, or traction).

   c. Horizontal displacement of one body on another more than 3.5 mm.

   d. Angulation greater than 11 degrees.

   e. Disruption of facets.

   f. Severe injury, such as multiple fractures at one segment.

D. Cervical spine fractures in children.
  1. Parameters for evaluation are different (see A, Normal and Abnormal Variants).
  2. Sites of involvement tend to be different: Teenagers have cervical injuries similar in distribution to adults and need to be evaluated with the full adult series of films. Children under 12 years old involve mostly the atlanto-occipital and atlantoaxial regions, so greatest attention should be paid to this region.

Thoracic Spine

A. Unusual site of fracture.
B. AP view: Watch interpediculate distance for abrupt widening suggestive of a burst fracture. There may be a paraspinous fullness secondary to hematoma. On chest films this may simulate adenopathy or large vessel bleed.
C. Lateral view: Swimmer's view is necessary for upper spine. Vertebral body height should remain uniform over the length of the thoracic spine (except for a normal slight anterior wedging of T_{12}). Normal kyphosis is 20 to 40 degrees, measured from T_4 to T_{12}.
D. Thoracic spine fractures generally are stable because of support from the thoracic cage and the orientation of the facets. Instability is more likely with multiple rib fractures or a sternal fracture.

Lumbar Spine

A. Normal appearance.
  1. On AP, interpediculate distance gradually widens from L_1 to L_5.
  2. On lateral, disk spaces gradually increase in size from L_{1-2} to L_{4-5}; L_5–S_1 may be slightly narrower.
  3. Limbus vertebra is an unfused ring apophysis secondary to anterior disk herniation that may simulate a fracture.

B. Patterns of injury[13]:
   a. Sixty percent of thoracolumbar fractures are at T_{12}–L_2.
   b. Ninety percent are at T_{11}–L_4.
   c. Seventy-five percent are compression fractures (anterior wedging or depression of the superior end-plate) with intact posterior elements.
   d. Twenty percent are fracture-dislocations (involvement of posterior elements as well as body). These are burst fractures, requiring CT to evaluate for bony fragments in the spinal canal and to choose anterior or posterior surgical approach. The facets may be fractured, subluxed, perched, dislocated, or locked, and these situations are not always obvious on CT unless the anatomy is understood.[14] Facets are identified by their orientation with respect to the vertebral body (superior facets are directed posteromedially and inferior facets are directed anterolaterally) as well as the shape of the articular surface (superior facet articular surface is concave, inferior articular surface is flat or convex; Fig 3–34). Using these guidelines, facet dislocation (AP or lateral or superior with "naked" facets) will not be missed on CT; subluxation of facets can be seen well with sagittal reconstruction.
   e. Chance fracture: Transverse fracture through the posterior elements (spinous process, pedicles, facets, transverse processes); the vertebral body may or may not be fractured. Lap-type seat belts act as a fulcrum. The tensile stress in hyperflexion causes the usual transverse fracture; little to no vertebral body compression; abdominal injury is often

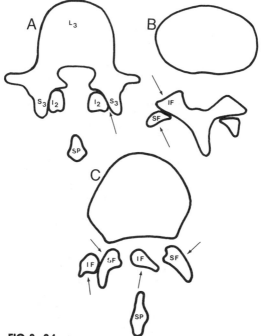

FIG 3–34.
CT appearance of facets in lumbar spine: **A,** normal. **B,** anterior lock. **C,** lateral lock. (Key: SF, superior facet; IF, inferior facet; SP, spinous process; *arrows,* articular surface)

associated.
C. Spondylosis: A defect in the pars intra-articularis, seen on the AP and lateral films, but often seen best on the oblique film (fracture through the neck of the "scotty dog").

1. Two thirds are at L_5.
2. More common in males than females.
3. More prevalent in whites than blacks.
4. Usually diagnosed in second or third decade.
5. Etiology uncertain but many feel that it is a stress fracture (repeated normal trauma), perhaps aggravated by dysplasia or hypoplasia of the pars.
6. May be unilateral or bilateral; if unilateral, hyperplasia and sclerosis of the contralateral facet may develop (simulating metastatic disease or osteoid osteoma); a congenitally absent pars may give a very similar appearance). If bilateral, associated spondylolisthesis may develop.

XII. NONACCIDENTAL TRAUMA (BATTERED CHILD SYNDROME)

> **Key Concepts:** Typical fractures or combinations of injuries are suggestive. Watch especially for fractures at different stages of healing, metaphyseal fractures, and rib fractures. Periosteal reaction and growth disturbances are also suggestive.

A. Radiographic signs considered to be pathognomonic:
1. Multiple fractures at different sites at different stages of healing (in the absence of metabolic or dysplastic disease).
2. Epiphyseal-metaphyseal fractures: Avulsion or bucket-handle fractures due to traction on the limb and avulsion of the metaphysis at the periosteal insertion (rare with accidental trauma in children under 5 years).
3. Rib fractures: Rare in accidental trauma owing to the plasticity of the thoracic cage. Fractures

are usually posterior, at costovertebral junction. Watch also for callus formation at the costochondral junction, simulating a rachitic or scorbutic rosary.

B. Unfortunately, many cases are not typical; other areas to watch for:

1. Extremity fractures (long and short tubular bones) are most frequent focus of injury, long bones much more common than hands or feet.

2. Periosteal reaction without definite evidence of fractures. (Differential diagnosis is extensive, including physiologic reaction, Caffey's disease, infection, tumor, rickets, scurvy, hypervitaminosis A, prostaglandin administration, normal wavy radius.)

3. Skull fractures, usually linear, are common.

4. Pelvic fractures are rare, usually pubic rami.

5. Clavicle, usually midshaft.

6. Spine: Rare, usually thoracic and lumbar flexion injuries.

7. Growth disturbances or traumatic metaphyseal cupping.

8. Extraskeletal injuries: Subdural hematoma, visceral injury, usually associated with skeletal injury.

C. Bone scan:

1. May be useful for additional or confirmatory evidence in specific cases.

2. Epiphyseal-metaphyseal fractures are difficult to detect because of the normally increased activity in this region.

3. Especially sensitive for occult rib, spine, or diaphyseal fractures, particularly acutely.

D. Suggested plain films for screening of clinically suspected cases:

1. AP of extremities (including hands, feet, and pelvis) and AP thorax.

2. AP and lateral skull.

E. Epidemiologic considerations.[15]

1. Overall frequency of skeletal trauma associated with abuse is low (20% to 40%).
2. Incidence of skeletal injuries considered pathognomonic for abuse is low.
3. Most common sites of fracture are diaphyses of long bones and skull, but these are not, by themselves, diagnostic.
4. In abused children skeletal trauma is most commonly seen in the first 2 years of life. This is especially true of skull fractures (90% are in this age group). These data suggest that the efficacy of skeletal surveys is decreased in older children. A selective approach, rather than skeletal survey, may be indicated for them.
5. Skeletal trauma is usually associated with clinical evidence of physical injury; it is much less commonly seen in conjunction with other forms of abuse (neglect or sexual).
6. Yield is extremely low in skeletal survey of siblings of abused children in the absence of clinical evidence of physical trauma.

XIII. MYOSITIS OSSIFICANS

> **Key Concepts:** The timing and zoning phenomenon of myositis is key to making the correct diagnosis. Without these considerations, osteosarcoma could be incorrectly diagnosed.

A. Juxtacortical myositis ossificans: Heterotopic formation of nonneoplastic bone and cartilage in soft tissue (usually muscle, but other soft tissue may be involved).
   1. Etiology: Usually traumatic, though the episode of trauma may be minor and not recalled.
   2. Sites: Anywhere, but most commonly areas prone to trauma (e.g., thighs, elbows).

3. Histologic evolution:
   a. Weeks 0–4: Pseudosarcomatous appearance of central zone suggests malignant neoplasm.
   b. Weeks 4–8: Centrifugal pattern of maturation: Periphery demarcated by initial immature osteoid formation, which organizes into mature bone about a cellular center (zone phenomenon).
4. Radiographic evolution:
   a. Weeks 0–2: Soft tissue mass (clinically painful, warm, doughy).
   b. Weeks 3–4: Flocculated amorphous densities within the mass, with periosteal reaction in underlying bone. At this stage, it may be mistaken for an early osteosarcoma since the calcification looks like tumor bone. Occasionally the mass may appear attached to underlying bone and may even elicit a periosteal reaction from the underlying bone, making it even more difficult to distinguish from tumor.
   c. Weeks 6–8: Sharp cortical bone surrounds a lacy pattern of new bone. Maturation proceeds centrifugally.
   d. Months 5–6: Maturity, with reduction in mass size. Often, a radiolucent zone separates the lesion from the underlying cortex.
5. The history and timing are crucial in supporting the early diagnosis of myositis ossificans. With good correlation of this information and radiographic findings, early and potentially confusing biopsies can be avoided.
6. Bone scans may be used serially to evaluate for maturation of myositis; surgical resection should not be considered prior to maturation, since it leads to a high rate of recurrence.
B. Posttraumatic periostitis: Broad-based bony mass

attached to bone resulting from trauma to the periosteum with hemorrhage beneath it.
   1. Evolution: Similar to that of myostitis ossificans.
   2. May have a "dotted veil" appearance at maturity.
C. Myositis ossificans associated with neurologic disorders or burns.
   1. Thirty-three to 49% of paraplegics show myositis in the paralyzed part, most commonly hips.
   2. Ossification in the muscles, tendons, and ligaments, not arising from the underlying bone.
D. Differential diagnoses of myositis:
   1. Parosteal osteosarcoma.
      a. Portions of the tumor may be separated from the underlying cortex by a radiolucent zone, but there is attachment; tends to wrap around long bone.
      b. Reversed zone phenomenon: More heavily calcified centrally, and the periphery is less dense and poorly circumscribed.
      c. Gradual increase in size.
   2. Periosteal osteosarcoma: Usually appears more aggressive, often with scalloping of underlying cortex and amorphous tumor bone formation.
   3. Juxtacortical chondroma: Scalloped underlying cortex, with juxtacortical calcific densities.
   4. Osteochondroma: Arises from the underlying bone, with continuation of cortical and medullary bone. Should not be confused with myositis.
   5. Tumoral calcinosis:
      a. Periarticular calcified soft tissue masses, usually hip, shoulder, and elbow that are entirely separate from the underlying bone, which is normal.

    b. Frequency higher among blacks; males and females affected in equal numbers.

    c. May progress to limit function, ulcerate, and undergo secondary infection.

    d. Very high recurrence rate after resection.

    e. Serum phosphate and ESR are elevated; normal calcium.

    f. Etiology may be enhanced renal tubular phosphate resorption.

    g. May respond to a program of phosphate binding antacids and dietary phosphate and calcium deprivation.

6. Myositis ossificans progressiva:

    a. Hereditary mesodermal disorder characterized by progressive ossification of striated muscles, tendons, and ligaments.

    b. May be autosomal-dominant with a wide range of expressivity; many spontaneous mutations.

    c. Target tissue thought to be interstitial tissues, with muscle involvement secondary to pressure atrophy.

    d. EMG studies have shown abnormalities consistent with a myopathy.

    e. Pathologic abnormalities are similar to those of myositis ossificans.

    f. Most frequent presenting symptom is acute torticollis with a painful mass in the sternocleidomastoid muscle.

    g. Progresses to the shoulder girdle, upper arms, spine and pelvis, with bridging between adjacent bones and, eventually, severe restriction of motion.

    h. High association of congenital digital anomalies: 75% show bilateral microdactyly of the first toes and/or synostosis of the phalanges; hallux valgus; thumbs less frequently involved.

    i. Remissions and exacerbations frequently are precipitated by minor trauma.

j. Involvement of the insertions of fasciae, ligaments, and tendons produce "exostoses."

XIV. ORTHOPEDIC HARDWARE

> *Key Concepts:* Femoral components of hip prostheses tend to loosen early, acetabular components late. Progression in signs of loosening is important to document. Subsidence of components may be subtle and is occasionally the only sign of failure. In knee prostheses, it is the tibial component that fails most often and commonly shows no radiographic abnormalities, even though there is clinical failure.

A. Total hip arthroplasty: Whether device is cemented or uncemented (with or without porous coating for bony ingrowth) the principles of radiographic analysis are similar:
  1. *Evaluation of placement*:
    a. *Lateral opening of the acetabulum* (measured by the angle of the cup to a line drawn between the ischial tuberosities) is *ideally 45 degrees* but should be no greater than 50 degrees. A wider opening angle increases the risk of dislocation.
    b. *Anteversion* of the acetabulum (evaluated on a lateral groin film) should be 10 to 15 degrees but may be as little as zero if compensated by an anteverted femoral component. Retroversion of the cup increases the probability of dislocation.
    c. Watch for *protrusio* acetabuli of the cup, especially in patients with RA, Paget's disease, or osteomalacia; lack of medialization of the cup may be a problem in OA.
    d. Evaluate for *lengths of gluteus medius and*

iliopsoas muscle groups; muscles are strongest at specific effective lengths. If they are overstretched, they may go into spasm, and if the space over which they contract is too short, they are ineffective. A side-to-side comparison of the position of the greater trochanters (insertion of gluteus) and lesser trochanters (insertion of the iliopsoas) should therefore be made. Adjustments may be made in the length of the neck of the femoral prosthesis or by transplanting the greater tuberosity more distally to achieve the effective length.

e. *Position of femoral component:* Neutral to slight valgus (prosthesis resting against the lateral cortex proximally and against the medial cortex distally) is preferred to varus (which predisposes to loosening).

2. *Evaluation for loosening:*

a. Malpositioning (see above) promotes loosening.

b. If cement is used for femoral component fixation, it must be placed in areas of maximum stress—around the tip, on the lateral side of the distal stem, and on the medial side of the stem proximally.

c. A fracture in cement definitely indicates loosening.

d. Lucency at the bone-cement interface is suspicious for loosening, but not diagnostic. This is especially true at the superolateral aspect of the acetabular component, where a 2- to 3-mm lucency is common. *To be diagnostic of component loosening, lucency at the bone-cement interface must show progression over time.*

e. Extensive scalloped resorption around a femoral component (*massive osteolysis*) *may be asymptomatic* and may be due to fibrous replacement *with little or no loosening.*

    f. Uncemented components often show a 1-
to 2-mm lucency around the component,
with a very thin sclerotic margin. In our
experience, unless this shows progression
or evidence of toggling (windshield wiper
sign) the lucency does not represent
loosening.

    g. Sclerosis at the tip of a femoral component
alone does not suggest loosening in
uncemented components.

    h. One of the most important signs of
loosening is *change in position of the
component. Acetabular components tend to
subside superiorly, and femoral components
tend to subside inferiorly*, with consequent
shortening of the leg. *Subsidence may be
extremely subtle*, since the bone-component
lucency may be obliterated by the change
in component position.

3. The pattern of loosening of hip prostheses is
interesting: Femoral components tend to
loosen early, with a rapid rate which then
falls off to a slower rate after 5 years. On the
other hand, the incidence of acetabular
component loosening is low initially and
shows a sharp increase after 5 years, so that
acetabular loosening approaches, and in some
series even surpasses, femoral component
loosening.

4. Infection is suspected with the presence of
periosteal reaction or large amounts of
heterotopic ossification. It is confirmed by
aspiration.

5. Bone scans show abnormal activity at the
surgical site of the prosthesis, which
decreases over a period of 9 months (longer
in uncemented prostheses). Increasing
activity over that period is abnormal.
Loosening generally shows focal activity at
the sites of greatest movement (especially the

tip), whereas infection shows a more general increase in activity.

6. Revision arthroplasties are often uncemented, and bone graft is used liberally to fill defects (very large pieces such as femoral heads may be used).
   a. Lateral "windows" in the femoral shaft may be performed for cement extraction in revisions.
   b. Shafts commonly fracture from prosthesis and cement extraction, reaming of thinned bone, and reinsertion of femoral components (especially long-stemmed components that do not conform to normal anterior femoral bowing). The fractures are often subtle vertical ones. Cerclage wiring protects the revision.
7. *Endoprostheses* are utilized when there is no need for an acetabular replacement (eg, in AVN or subcapital fractures). Either the older Austin Moore or Thompson prosthesis with a large head is used, or a bipolar (or Bateman) component (a cup that clamps on to the head of the femoral component; the femoral head can move within the cup, and the cup can move within the patient's normal acetabulum).

B. Total knee arthroplasty.
1. Normal placement:
   a. Tibial component is placed 90 degrees ± 5 degrees to the long axis of the tibial shaft on the AP and ranges from 90 degrees to the long axis of the tibia to a slight posterior tilt on the lateral.
   b. Femoral component is placed in 5 degrees' ± 5 degrees valgus on the AP and 90 degrees' ± 5 degrees to the long axis of the femoral shaft on the lateral.
   c. Widened joint space on one side suggests instability.

2. Evaluation of loosening:
   a. As in the hip, watch for progressive lucency at bone-cement or bone-uncemented component interface.
   b. Subsidence or other change in component position indicates loosening.
   c. The tibial component is much more likely to loosen than is the femoral one.
   d. Loosening and/or infection in knee components is much more difficult to detect radiographically than in hip arthroplasties. Technical factors such as slight flexion, rotation, or change in centering can mask a lucency at the bone-component interface. To avoid those technical difficulties, fluoroscopic spot films of the tibial component-bone interface should be taken in order to evaluate for progressive lucency.
   e. In uncemented knee components it is common to see a nonprogressive 1- to 2-mm lucency at the tibial component-bone interface, which may have a sclerotic margin. It is speculated that the lucency may represent fibrous union. With porous-coated prostheses, there may even be loosened microspheres in this lucent region. As long as there is no progression, these findings are not likely to indicate loosening.
3. The femoral component rarely shows loosening. Uncemented femoral components frequently show the effect of stress shielding, with bone resorption (but no sclerotic margin) at parts of the bone-prosthesis interface. Buttressing may be seen at the sites of stress (often the anterior cortex).
4. Patellar complications include stress fracture and superior subsidence of the patellar button.

C. Dynamic screws (Richard's screw is one common variety) for intertrochanteric fractures.
   1. Reduction may be stable and acceptable if it is anatomic or has had medial displacement of the shaft and impaction or valgus repositioning.
   2. The screw fits within a coaxial sleeve in the femoral neck and is designed to "telescope" into its sleeve several centimeters to avoid cutting out of an osteoporotic femoral head if there is collapse or impaction at the fracture site.
   3. The position of the screw head should be slightly inferior and posterior to the middle of the femoral head and very close to the articular surface.
D. External fixators.
   1. Used to maintain the length of a fractured limb in the presence of infection or severe comminution.
   2. Rotation must be watched for carefully.
E. Cerclage wires or Parham bands (older, wider, rarely used today) may be placed around comminuted shaft fractures but may delay healing by stripping the periosteum and devascularizing the bone.
F. Intermedullary rod.
   1. Watch for shaft shortening if there is a butterfly fragment present.
   2. Rotation should be watched for, especially if the fracture is beyond the isthmus (generally proximal or distal third of long bones), so that the rod does not occupy the entire canal of the fractured segments. Rotation may be controlled by placing interlocking nails or rods.
   3. Significant diastasis at the fracture site must be avoided.

G. Swanson arthroplasties (MCP, MTP, IP joints).
   1. Hinge is thinnest portion and, therefore, at risk for fracture.
   2. Dislocation of the flange may occur, especially in diseases such as RA with soft tissue imbalance or contractures.
H. Carpal implants.
   1. Dislocation and rotation both are common.
   2. Reactive synovitis and erosive changes may be seen.
I. Posterior spinal instrumentation: See Chapter 5, Section I. Scoliosis.
J. Anterior spinal instrumentation.
   1. In the cervical spine, complications arise from breakage or backing out of screws or from extrusion of the bone plug placed in the intervertebral disc space.
   2. In the lumbar spine, hardware failure most commonly involves screw breakage. Screws may cut out through osteoporotic bone. Hardware placed on the left side is associated with aortic complications.

REFERENCES

1. Fisher M, Rogers L, Hendrix R, et al.: Carpometacarpal dislocations. *CRC Crit Rev Diagnostic Imaging* 1984; 22: 95.
2. Dunn A: Fractures and dislocations of the carpus. *Surg Clin North Am* 1972; 52:1513.
3. Gilula L: Carpal injuries: Analytic approach and case exercises. *AJR* 1979; 133:503.
4. Yeager B, Dalinka M: Radiology of trauma to the wrist: Dislocations, fracture dislocations, and instability patterns. *Skel Radiol* 1985; 13:120–130.
5. Manaster B: Digital arthrography of the wrist. *AJR* 1986; 147:563–566.
6. Crowley D, Reckling F: Supracondylar fracture of the humerus in children. *Ann Fam Physician* 1972; 5:113.

7. Deutsch A, Resnick D, Mink J, et al.: Computed and conventional arthrotomography of the gleno-humeral joint: Normal anatomy and clinical experience. *Radiology* 1984;153:603–609.

8. Cone R, Resnick D, Danzing L: Shoulder impingement syndrome: Radiographic evaluation. *Radiology* 1984; 150:29-33.

9. Gill K, Bucholy R: The role of CT scanning in the evaluation of major pelvic fractures. *J Bone Joint Surg Am* 1984; 66A:34.

10. Affram P: An epidemiologic study of cervical and trochanteric fractures of the femur in an urban population. Analysis of 1664 cases with special reference to etiologic factors. *Acta Orthop Scand Suppl* 1964; 64:11.

11. Bayliss A, Davidson J: Traumatic osteonecrosis of the femoral head following intracapsular fracture: Incidence and earliest radiological features. *Clin Radiol* 1977; 28:407.

12. Laurin C, Dussault R, Levesque H: The tangential xray investigation of the patellofemoral joint. *Clin Orthop* 1979; 144:16.

13. Nicoll E: Fractures of the dorso-lumbar spine. *J Bone Joint Surg Br* 1949; 31:376.

14. Manaster B, Osborn A: CT patterns of thoracolumbar facet dislocation. *AJR* 1987; 148:335-340.

15. Merton D, Radkowski M, Leonidas J: The abused child: A radiological reappraisal. *Radiology* 1983; 146:377–381.

BIBLIOGRAPHY

Keats T: *An Atlas of Normal Roentgen Variants That May Simulate Disease.* Chicago, Year Book Medical Publishers, 1980.

Rang M: *Children's Fractures.* Philadelphia, JB Lippincott, 1974.

Rockwood C, Green D: *Fractures in Adults.* Philadelphia, JB Lippincott, 1984.

Rogers L: *Radiology of Skeletal Trauma,* vols 1 and 2. New York, Churchill Livingstone, 1982.

4

Metabolic Bone Disease

I. OSTEOPOROSIS

> **Key Concepts:** Abnormally decreased volume of bone which is, however, normal histologically and in its degree of mineralization; senile generalized osteoporosis, the most common form, contributes to significant morbidity, especially in older females; osteoporosis in childhood has a long and interesting differential diagnosis.

A. Definition: Any condition in which there is a reduction in the amount of bone tissue.
   1. This is a specific term and is not synonymous with osteopenia or deossification.
   2. Any bone tissue that is present is fully mineralized and normal histologically.
   3. Lab values are normal.
   4. Other major causes of osteopenia may be difficult to differentiate from osteoporosis.
      a. Osteomalacia: Looser's lines may be present; also, the trabeculae may appear smudged or indistinct.
      b. Hyperparathyroidism: The presence of

subperiosteal resorption, brown tumors, or end-plate sclerosis ("rugger jersey spine") is helpful.

c. Multiple myeloma: Focal, punched-out lesions help make the diagnosis, but generalized myeloma may be indistinguishable from osteoporosis.

B. Etiologies and radiographic findings: Generalized osteoporosis.

1. Senile (involutional) osteoporosis.[1]

a. In women, loss of bone mass begins before menopause (perhaps even in the late childbearing years) and accelerates much more rapidly than in men. The time of onset and rate of acceleration determine the risk for fracture.

b. Incidence is higher among women than men and among Caucasians and Asians than blacks.

c. Modified by estrogens, calcium intake, weight (thin women store less estrogen), weight-bearing exercise, family history, amenorrhea (as in athletes or after oophorectomy), smoking, heavy alcohol use.

d. Thirty to 50% of women over 60 show evidence of significant bone loss.

e. The process in usually irreversible, but the rate of bone loss may be altered by therapy.

f. Etiology is unclear. Some studies favor a decrease in matrix formation, while others suggest increased osteoclast activity.

g. Trabecular bone is resorbed faster than cortical bone.

h. Most significant sites include the spine, femur (subcapital neck and intertrochanteric region), distal radius, and proximal humerus.

i. Bone loss of 30% to 50% is required before

it can reliably be detected by plain film.

j. The spine shows the earliest radiographic signs: There is a loss of density due to resorption of the horizontal trabeculae. The vertical trabeculae appear more distinct and prominent. There is cortical thinning and only a relative increase in density of the end-plates (there is no significant subchondral sclerosis).

k. Compression fractures are common and repair is slow. Fractures may be in the form of anterior wedging, biconcavity of the bodies, or true compression.

l. The generalized osteoporosis is uniform and lacks cortical striation or tunneling (which indicate faster remodeling).

m. In the femoral neck, the major stress-bearing trabeculae are accentuated.

n. With fluoride therapy, trabecular accentuation and stress fractures occur.

2. Hypercorticism (Cushing's disease or exogenous steroids) differs from osteoporosis in the following ways:

a. Sclerotic (thickened) vertebral end-plates (callus from microfractures); abundant callus formation elsewhere as well.

b. Aseptic necrosis, especially femoral and humeral heads, is a complication.

c. Diaphyseal infarcts.

d. Occasional Charcot-like joints.

3. Homocystinuria: Young patient with scoliosis, biconcave vertebrae, and arachnodactyly.

4. Acromegaly: Should have typical hand findings of increased cartilage and soft tissues, bony excrescences, and spadelike tufts.

5. Ochronosis: See Chapter 2.

6. Amyloidosis: See Chapter 2.

7. Hyperthyroidism.

8. Alcoholism: Direct effect on osteoblasts.

9. Osteogenesis imperfecta: Discussed in a later chapter but characterized by diaphyseal fracture, bowing of long bones, and blue sclerae.
10. Drug-related:
    a. Heparin (requires daily doses in excess of 15,000 units).
    b. Phenobarbital.
    c. Dilantin.
    d. Steroids.
11. Mastocytosis: See IX, below.
12. Idiopathic juvenile osteoporosis: Acute onset in a previously healthy 8- to 12-year-old; osteoporosis with fracture and loss of height of vertebral bodies. Metaphyseal fractures are common; eventually stabilizes, but the osteoporosis remains; diagnosis of exclusion.

C. Localized osteoporosis.
   1. Disuse: May be uniform, spotty, or very aggressive-looking with cortical tunneling and metaphyseal bandlike lucencies.
   2. Reflex sympathetic dystrophy (Sudeck's atrophy).
      a. Soft tissue trophic changes (swollen, with hyperesthesia, then atrophic and contracted).
      b. Severe extremity involvement distal to the affected site: rapid severe osteoporosis with a moth-eaten pattern, cortical tunneling, and lucent metaphyseal bands.
      c. Mediated by the sympathetic nervous system.
      d. Precipitating cause may not be identified but may be virtually any musculoskeletal, neurologic, or vascular condition.
   3. Transient regional osteoporosis.
      a. Self-limited.
      b. Cartilage remains intact, differentiating it from septic arthritis.

  c. Two forms:
    (1) Migratory: Especially knee, ankle, and foot; male predominance; onset in fourth or fifth decade. Local pain, swelling, and osteoporosis; resolves within 1 year; recurs around other joints.
    (2) Transient osteoporosis of the hip: described in a woman in the third trimester of pregnancy involving the left hip; in males, either hip may be involved; resolves within 1 year.
D. Quantitative analysis of bone mass.
  1. Quantitative computed tomography (QCT).
    a. Single-energy gives approximately 97% accuracy.
    b. Dual-energy not practical in most departments but may be more accurate.
    c. Radiation approximately 300 mR.
    d. Fat content in spine adversely affects accuracy, making this generally less accurate in elderly women.
    e. Positioning must be accurate for reproducible results and serial studies.
    f. Major advantage is that only the trabecular bone is imaged, which is thought to be the area of interest in bone mass studies.
  2. Dual-photon absorptiometry: Transmission scanning with an isotope source (^{153}Gd).
    a. Accuracy similar to that of QCT.
    b. Significantly less radiation.
    c. Major disadvantage is that all the bone (cortical as well as trabecular) is imaged. Aortic calcification may also be a factor. It may also be difficult to distinguish a compression fracture, thus giving a false impression of increased bone mass due to impaction.

II. RICKETS/OSTEOMALACIA

> ***Key Concepts:*** Lack of mineralization of normal osteoid. Rickets is characterized by a widened zone of provisional calcification and metaphyseal irregularity; osteomalacia is characterized by coarsened trabeculae and Looser's lines.

A. Definition:
   1. Lack of mineralization in osteoid that is otherwise normal.
   2. Laboratory values (Fig 4–1):

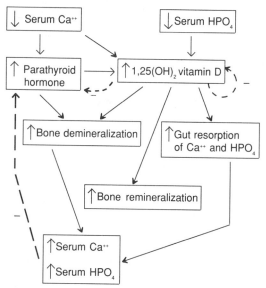

FIG 4–1.
Algorithm of laboratory findings in rickets and osteomalacia.

        a. Serum: decreased Ca^{++} and HPO_4; elevated alkaline phosphatase.

        b. Urine: decreased Ca^{++} and HPO_4.

B. Radiographic signs:

  1. Generalized osteopenia due to decreased number of trabeculae.

  2. Coarsened trabeculae due to osteoid deposition on remaining trabeculae without adequate mineralization.

  3. Looser's lines: Wide transverse lucencies, often incomplete. Unmineralized osteoid is deposited during the repair process. Characteristic sites, often symmetric—pubic rami, ribs, axillary margins of scapulae, medial margins of proximal femurs, posterior margin of proximal ulna.

  4. In rickets, changes predominate at sites of rapid growth: Proximal humerus, distal radius, distal femur, both ends of tibia. A cost-effective survey would, therefore, include anteroposterior (AP) view of knees, wrists, and ankles.

  5. Rickets abnormalities are predominantly metaphyseal, with a widened lucent zone of provisional calcification consisting of unmineralized osteoid. The metaphyses become widened, irregular, and cupped, and occasionally the epiphyses slip owing to normal weight-bearing stresses on an abnormally fragile epiphyseal plate.

  6. Bearing weight may also produce bowing deformities of the long bones and vertebral end-plate depression.

C. Metabolism:

  1. Vitamin D (cholecalciferol): A prohormone with two sources, it is produced endogenously in the skin (activated by ultraviolet light) and derived exogenously from dairy products and other food supplements.

2. Vitamin D undergoes two hydroxylations to become an active metabolite:
   a. In liver, 25 (OH) vitamin D.
   b. In kidney, 1,25 $(OH)_2$ vitamin D.
3. Vitamin D function:
   a. Maintains normal serum Ca^{++} and HPO_4.
   b. Maintains bone mineralization.
4. Effects of vitamin D on organ systems:
   a. Bone: Mobilizes Ca^{++} and HPO_4 by stimulating osteoclasts (requires parathyroid hormone [PTH]). By maintaining normal serum Ca^{++} and HPO_4, it maintains bone mineralization; also stimulates bone mineralization directly (independent of serum Ca^{++} levels).
   b. Gut: Stimulates increased resorption of Ca^{++} and HPO_4.
   c. Kidney: Stimulates tubular resorption of HPO_4.
   d. Parathyroid gland: Suppresses PTH (independent of serum Ca^{++} levels).
5. Regulation of vitamin D:
   a. Increased by low dietary intake of Ca^{++} (normally mediated by PTH, but may also regulate independently).
   b. Increased by hypophosphatemia (direct effect, not mediated by PTH).
   c. Decreased by vitamin D (self-regulator) directly as well as indirectly through decreased PTH.

D. Etiologies of osteomalacia and rickets (note that radiographically these cannot be differentiated reliably:
   1. Dietary calcium deprivation (rare in United States).
   2. GI malabsorption (due to surgery or various malabsorption syndromes).
   3. Liver abnormalities:
      a. Hepatocellular disease interferes with

hydroxylation of the prohormone vitamin
D.

   b. Biliary disease relates to malabsorption of
vitamin D.

4. Drug therapy: Phenobarbitol, Dilantin,
Didronyl.

5. Renal osteodystrophy: Combination of
osteomalacia (lack of second hydroxylation
of vitamin D) and secondary
hyperparathyroidism (see Chapter 4,
Section III).

6. Renal tubular disorders:

   a. Also known as vitamin D-resistant (or
refractory) rickets.

   b. Several syndromes with renal tubular
defects, including cystinosis and X-linked
hypophosphatemia (males with mild
rachitic changes and only mild
osteopenia).

   c. Rickets is due to renal tubular phosphate
loss.

   d. The laboratory values are therefore
different from classic rickets: decreased
serum Ca^{++} and serum HPO_4, and
elevated urine Ca^{++} and urine HPO_4.

7. Tumor-associated rickets (also vitamin D
refractory):

   a. Autosomal-dominant.

   b. Associated with small benign soft tissue or
skeletal tumors (hemangioma,
nonossifying fibroma, giant cell tumor).

   c. Possible tumor-related humoral substance
causing decreased renal tubular phosphate
resorption.

   d. Normocalcemic; decreased serum HPO_4;
elevated urine HPO_4 and alkaline
phosphatase.

   e. Responds to tumor resection.

8. Neonatal rickets:

   a. Rickets normally is not seen

       radiographically before 6 months of age.
- b. May be seen in premature infants with low birth weight and long periods of parenteral nutrition.
- c. Due to rapid skeletal growth and possible insufficiency in dietary calcium, phosphate, and vitamin D.
- d. May be complicated by liver or kidney disease.
- e. Copper deficiency in premature infants mimics rickets.

E. Other etiologies that mimic rickets:
  1. Hypophosphatasia:
     - a. Severely decreased alkaline phosphatase.
     - b. Elevated phosphoethanolamine in blood and urine.
     - c. Wide spectrum of disease.
     - d. Severe: Victims die in infancy; "tar babies"; no cranial mineralization (only face and base of skull); short bowed bones with severe fractures and deformities as well as ricketslike changes at metaphyses.
     - e. Less severe: Osteopenic; looks like rickets at metaphyses but there is craniostenosis and wormian bones; improves progressively.
     - f. Least severe: Delay in fracture healing and mild bowing deformities.
  2. Metaphyseal chondrodysplasia (Schmid type):
     - a. Abnormal endochondral bone formation.
     - b. Normal serum values.
     - c. Short bowed bones with widened growth plates; bony projections from the metaphysis into the growth plate.
     - d. *Normal* mineralization and bone density.

III. HYPERPARATHYROIDISM (HPTH)

> ***Key Concepts:*** Osteopenia or osteosclerosis; soft tissue and vascular calcification, several varieties of resorption, brown tumors, chondrocalcinosis. Though some manifestations are more prominent in primary than secondary HPTH, all may be found in either.

A. Laboratory values: Elevated serum Ca^{++} and alkaline phosphatase, decreased serum HPO_4; elevated urine HPO_4 and Ca^{++}.

B. Etiology:
   1. Primary HPTH: 90% due to adenoma.
   2. Secondary HPTH: Due to renal dysfunction; often see findings of rickets or osteomalacia as well.

C. Clinical findings:
   1. Nephrolithiasis.
   2. Gastric or duodenal ulcers.
   3. Weakness due to hypocalcemia.
   4. Bone pain and tenderness (10% to 25%).

D. Radiographic findings:
   1. Radiographically identifiable skeletal manifestations in 50%.
   2. For screening and serial studies, posteroanterior (PA) view of hand is most cost-effective.
      a. Predilection of many sites in the hand for manifestations of HPTH (subperiosteal resorption on radial side of digits, tuft resorption, soft tissue calcification, chondrocalcinosis of triangular fibrocartilage complex).
      b. Thin soft tissues and small bones, so bone detail is seen better.
      c. Fine detail extremity filming is performed routinely.
   3. Skeletal manifestations (Table 4–1):

TABLE 4–1.
Skeletal Manifestations of Primary and Secondary HPTH

Manifestations	Most Common in	
	Primary HPTH	Secondary HPTH
Generalized osteopenia	+	+
Resorption	+	+
Sclerosis		+
Brown tumor	+	
Chondrocalcinosis	+	
Soft tissue calcification		+
Vascular calcification		+

a. Generalized osteopenia; HPTH results in either excessive bone resorption or, occasionally, bone formation. There is a great variation in these activities, resulting in a spectrum of appearances, but generalized osteopenia is most common.

b. Resorption; very common manifestation in both primary and secondary HPTH.

   (1) Subperiosteal resorption: Most common form, affects the radial aspect of phalanges (especially middle phalanx of index and middle finger); also affects the tufts of fingers and margins of phalanges, giving marginal erosions; seen at medial proximal humeral, femoral, and tibial shaft.

   (2) Endosteal resorption: Scalloped defects; usually not an isolated finding.

   (3) Intracortical resorption: Linear striations and tunneling; not a specific finding (seen in any manifestation of rapid bone resorption).

   (4) Trabecular resorption: Especially well seen in the diploe of the cranium, resulting in salt-and-pepper skull.

   (5) Subchondral resorption: Results in collapse of subchondral surface and apparent widening and sclerosis of joint. Sacroiliac, acromioclavicular, temporomandibular joints and symphysis pubis are most common sites; must differentiate from sacroiliitis and other arthritides.

   (6) Subligamentous resorption: At sites of ligament origins, especially on pelvis, calcaneus, and distal clavical inferiorly.

c. Sclerosis: May be patchy or diffuse and is much more common in secondary than in primary disease. Mechanism is unclear, but parathyroid hormone (as well as vitamin D) is known to stimulate

osteoblastic activity. Bone formation may be more dramatic than bone resorption in some patients; this, combined with an element of subchondral resorption, collapse, and impaction, produces the rugger jersey appearance of dense vertebral end-plates seen most commonly in secondary HPTH.
d. Brown tumors: Localized accumulations of osteoclasts producing expanded lytic lesions. Fibrous tissue and giant cells are also present. Radiographically and pathologically it may be difficult to differentiate from giant cell tumor. Usually other manifestations of HPTH are present as well, making the diagnosis. More common in primary than in secondary HPTH. May hyperossify after resection of the adenoma, giving the appearance of blastic metastases.
e. Chondrocalcinosis: More common in primary than in secondary HPTH.
f. Periarticular and soft tissue calcifications are more common in secondary than in primary disease. Vascular calcification may also be prominent. Large "tumoral" deposits most commonly are seen in dialysis patients.

IV. RENAL OSTEODYSTROPHY

Key Concepts: Chronic renal failure with a combination of bone abnormalities due to osteomalacia and secondary HPTH—and aluminum intoxication if the patient is on dialysis. The manifestations are highly individual, depending on the predominant disease in each patient.

A. Definition:

1. Bone disease seen in patients with chronic renal failure.
2. Main features are osteomalacia or rickets, secondary HPTH and/or aluminum intoxication (if patient is on dialysis).
3. In young patients, expressed mostly as rickets, with later changes of HPTH. More mature patients have more marked changes of HPTH.
4. Secondary HPTH is due to renal failure which causes phosphate retention; phosphate retention causes a decreased serum calcium level, which in turn causes parathyroid hyperplasia. In renal failure, there is also decreased calcium resorption from the gut as well as skeletal resistance to PTH.
5. Aluminum intoxication:
   a. Sources are aluminum salts in the dialysis fluid and aluminum hydroxide antacids administered to control hyperphosphatemia.
   b. Radiographic features are osteomalacia and stress fractures; resembles rickets in children.
   c. Clinical symptoms are bone pain, proximal muscle weakness; may be associated with dialysis encephalopathy.
   d. No response to vitamin D therapy.
   e. Distinctive histology: Osteomalacia with aluminum deposits at the interface of mineralized bone and osteoid.
B. Laboratory values: Serum Ca^{++} normal to decreased, serum phosphate and alkaline phosphatase elevated.
C. Radiographic features:
   1. *Rickets/osteomalacia:* Features osteopenia, widened zone of provisional calcification, and Looser's lines. *Slipped epiphyses* not uncommon in chronic disease.
   2. *Secondary HPTH: Resorption, osteosclerosis,* and

soft tissue and vascular calcifications are prominent features. Chondrocalcinosis and brown tumors are less commonly seen. With severe chronic disease, *periosteal neostosis* may be seen, most commonly in the femurs, pubic rami, and metatarsals. There may be a lucent zone between the periosteal new bone and host bone. If present, neostosis helps distinguish secondary from primary HPTH.

3. Abnormalities following dialysis:
   a. Changes of HPTH often resolve (depending on amount of calcium in the dialysis fluid and control of phosphate): reversal of resorption, healing brown tumors, resolution of vascular and soft tissue calcification.
   b. Effect on rickets/osteomalacia is controversial.
   c. Soft tissue calcification is common (50%) and may occur in large "tumoral" deposits.
   d. Osteomyelitis and septic arthritis (arteriovenous [AV] shunt is port of entry; patients often are chronically immunosuppressed).
   e. Aseptic necrosis: Due to steroid therapy as well as subchondral resorption and collapse.
   f. "Dialysis cysts" in phalanges and carpus; significance unknown.
   g. Shunt aneurysm.
   h. Aluminum intoxication: Osteomalacia and stress fractures.
   i. Carpal tunnel syndrome: Possibly related to altered hemodynamics at AV shunt.
   j. Spondyloarthropathy: Disc space narrowing, end-plate erosions and sclerosis, no osteophyte formation; looks like infection; biopsy should be taken since these patients are at risk for

infection, but some material should also be evaluated for crystal deposition and amyloid.

V. HYPOPARATHYROIDISM

> *Key Concepts:* Osteosclerosis or -porosis and soft tissue calcification are common to all varieties; pseudo- and pseudopseudohypoparathyroidism also have a characteristic somatotype, shortened metacarpals or metatarsals, and small exostoses.

A. Definition: Deficiency in PTH production, usually secondary to excision or surgical trauma; rarely idiopathic; patient may have hypocalcemia and neuromuscular symptoms.

B. Radiographic abnormalities:
   1. *Osteosclerosis*, generalized or localized, is the most common finding.
   2. *Hypoplastic dentition.*
   3. *Subcutaneous calcification.*
   4. *Basal ganglia calcification.*
   5. Less common abnormalities:
      a. Osteoporosis.
      b. Premature closure of epiphyses.
      c. Enthesopathy and vertebral hyperostosis.
      d. Bandlike areas of increased density in metaphyses and end-plates of vertebral bodies.

C. Pseudohypoparathyroidism:
   1. Etiology: End-organ resistance to PTH.
   2. Characteristic somatotype: Short, obese, brachydactyly.
   3. Has features similar to those of hypoparathyroidism:
      a. Hypocalcemia.
      b. *Osteosclerosis.*
      c. *Basal ganglia calcifications.*

       d. *Soft tissue calcifications.*
4. Features that distinguish it from hypoparathyroidism.
       a. *Short metacarpals, metatarsals, and phalanges,* especially first, fourth, and fifth.
       b. Small *exostoses* projecting at right angles from the bone.
       c. Growth deformities, with wide bones and coned epiphyses.
D. Pseudopseudohypoparathyroidism:
    1. Normocalcemic form of pseudohypoparathyroidism with end-organ resistance to PTH.
    2. Somatotype identical to that of pseudohypoparathyroidism.
    3. Radiographic abnormalities identical to those of pseudohypoparathyroidism except that basal ganglia calcifications are rare.

VI. THYROID DISEASE

A. Juvenile hypothyroidism
    1. *Delayed skeletal maturation,* may be severe.
    2. *Wormian bones.*
    3. Epiphyseal dysgenesis.
       a. *Stippled epiphyses* in infancy.
       b. *Fragmented epiphyses in childhood, especially the proximal femoral capital epiphysis.* This condition, which is called cretinoid epiphyses, may be misdiagnosed as Legg-Calvé-Perthes disease.
       c. *Coned epiphyses* may occur later.
    4. May have short vertebral bodies at the junction of the thoracic and lumbar spine, with anterior beaking and kyphosis (nonspecific).
    5. The diagnosis must be picked up in infancy, when it can be reversed in order to prevent mental retardation.
B. Hyperthyroidism:

1. In children, accelerated skeletal maturation.
2. In adults, *increased bone turnover and consequent osteoporosis*, with typical osteoporotic vertebral body pathologic fractures and kyphosis.
3. *Myopathy* may simulate arthritis.

C. *Thyroid Acropachy.*

A rare complication following therapy for hyperthyroidism; the patient may be euthyroid, hypothyroid, even occasionally hyperthyroid.

1. The *metacarpals and phalanges* are most commonly involved, *often asymmetrically*; other long bones involved only occasionally.
2. Most common feature is a *dense feathery periosteal reaction.*
3. *Soft tissue swelling.*
4. *Clubbed fingers.*

VII. ACROMEGALY

A. Definition: *Excess of growth hormone* produces *proportional increase in size in the skeletally immature patient and tubular bone widening and acral growth in the skeletally mature patient.*

B. Radiographic abnormalities:

1. *Soft tissue thickening, especially over the phalanges and in the heel pad.* (The distance from the calcaneus to the plantar aspect of the heel is normally less than 23 mm in males and 21.5 mm in females. The wide range of normal makes the question of heel pad thickness controversial.)
2. *Enlarged sella*, often with destructive changes.
3. *Prominent facial bones* and occipital protuberance, with *enlarged, excessively pneumatized sinuses.*
4. Increased vertebral body and disc height; posterior vertebral scalloping; exaggerated thoracic kyphosis.
5. *Hand and foot changes predominate over more proximal bones*; the bones are *wide*, with

spadelike tufts. Excresences at tendon attachments along the phalanges are prominent. Overgrowth causes *cartilage widening,* and beaking osteophytes eventually develop into secondary degenerative joint disease (DJD).
6. There is also bony proliferation at the entheses.
7. Osteoporosis may be a late feature.

VIII. SCURVY (HYPOVITAMINOSIS C)

A. Definition: Lack of vitamin C results in *abnormal collagen formation;* hemorrhage is frequent, and bone production is decreased.
B. Radiographic abnormalities in children:
1. Rarely seen before 6 months of age.
2. *Subperiosteal hemorrhage* may cause spectacular elevation of the loosely attached periosteum in children, resulting in *extensive subperiosteal bone formation.*
3. The abnormal bone production is most manifest at sites of rapid growth.
   a. Costochondral junction *(scorbutic rosary).*
   b. *Ends of long bones* (especially around the knee).
      (1) *Disorganized growth zone becomes densely calcified,* giving the *sclerotic epiphyseal rim* (Wimburger's sign) and the *dense metaphyseal line* (white line of Frankel).
      (2) More proximal metaphyseal lucent line.
      (3) With minor trauma, this brittle bone develops *metaphyseal corner fractures* (Pelkin's fracture).
      (4) *Generalized osteopenia.*
C. Radiographic abnormalities in adults: Nonspecific osteopenia and resultant pathologic fracture.

IX. MASTOCYTOSIS

A. Definition: *A proliferative disorder of mast cells* resulting in bone abnormalities as well as the

clinical symptoms of flushing, nausea, vomiting, and skin lesions resembling urticaria pigmentosa.
B. Radiographic abnormalities: May show *either* osteoporosis or sclerosis.
   1. *Histamine release* from mast cells *results in osteoporosis*, usually generalized but occasionally focal.
   2. *Host reaction* to marrow infiltration *may result in bony sclerosis*, usually focal but occasionally diffuse.

X. GAUCHER'S DISEASE

> **Key Concepts:** Storage disease: hepatospleno-megaly, femoral head avascular necrosis (AVN), osteoporosis, Erlenmeyer flask deformity of the knee.

A. Definition: Sphingolipid storage disorder, with accumulation in the reticuloendothelial system and marrow infiltration with Gaucher cells. Life span is normal in the usual form, but there are infantile and juvenile forms that produce mental retardation and early death.
B. Epidemiology:
   1. Familial; no sex predilection.
   2. Onset in childhood or young adulthood.
   3. Many patients are Ashkenazi Jews.
C. Radiographic abnormalities:
   1. *Erlenmeyer flask deformity* (expansion of the distal femur): Due to marrow infiltration, present in 40% to 50% of patients.
   2. *AVN*, especially of femoral head present in 40% to 50% of patients.
   3. *Generalized osteoporosis* with fractures, especially of vertebral end-plates (H-shaped, as in sickle cell disease).
   4. Bone infarction with focal sclerosis and occasional bone-within-bone appearance.

5. Occasional focal lytic "cystic" areas, simulating neoplasm.
6. *Hepatosplenomegaly.*
7. Increased susceptibility to osteomyelitis.
D. *Differential diagnoses for Erlenmeyer flask* deformity:
1. *Anemias*: May appear very similar in that there may also be H-shaped vertebral end-plate collapse, AVN, and bony infarcts. Splenomegaly in Gaucher's should help differentiate.
2. *Niemann-Pick disease*: Sphingomyelin accumulation with similar infiltrative bony findings (except AVN) as well as hepatosplenomegaly.
3. *Pyle's disease*: A metaphyseal dysplasia that results in expanded metaphyses of tubular bones (especially about the knee) and normal diaphyses.

XI. MYELOFIBROSIS

A. Definition: *Fibrosis in areas of the skeleton normally involved in hematopoesis,* with subsequent compensatory hematopoesis in the fatty marrow of the large tubular bones. These latter sites, in turn, may become fibrotic.
B. Radiographic findings:
1. *Sclerotic bone marrow* (or patchy increased density with cortical thickening) in the hematopoetic bones *(vertebrae, pelvis, and ribs)* as well as long tubular bones.
2. Extramedullary hematopoesis: Hepatosplenomegaly and paraspinous masses.
3. Occasional periosteal reaction secondary to bleeding associated with platelet deficiencies.
C. Differential diagnoses:
1. Mastocytosis, fluorosis, Paget's disease, metastases.
2. Splenomegaly and paraspinous masses help to differentiate.

XII. PAGET'S DISEASE OF BONE

> *Key Concepts:* Skull, vertebral bodies, and pelvis most commonly affected; three sequential stages—lytic, mixed, sclerotic; affected bone enlarged; skeletal deformities, pathologic fracture, and occasional sarcomatous change; elevated alkaline phosphatase.

A. Definition: A disease of unknown etiology that causes abnormal remodeling of bone with *simultaneous osteoclast and osteoblast activity.* Thickened, disorganized, *fragile trabeculae* result.

B. Epidemiology:
   1. *Males* more commonly affected than females.
   2. *Rare before age 40 years.*
   3. Most common in Great Britain, Australia, and the United States; rare in Asia.

C. Laboratory abnormalities:
   1. *Elevated serum alkaline phosphatase* (due to bone formation).
   2. Elevated serum and urinary hydroxyproline (due to bone resorption).
   3. These values vary with the severity of the disease and are often normal initially.

D. Clinical signs:
   1. Often asymptomatic.
   2. Local pain, bowing of long bones, or pathologic fracture.
   3. Enlarging bone, especially cranium.
   4. Deafness (due to either middle ear ossicle involvement or cranial nerve compression).
   5. Occasional spinal cord signs from platybasia or other compression.
   6. Rare CHF secondary to high output from increased blood flow in involved bone.

E. Radiographic appearance: *three sequential stages* (though they may coexist):
   1. *Lytic*: Initial destruction, usually geographic

with a well-defined border (flame-shaped border in tubular bones).

2. *Mixed*: Intermediate stage, with cortical accretion and enlarging bones.
3. *Sclerotic*: Continued increase in both density and size.

F. Most common sites of involvement:

1. *Spine (75% of patients)*: Lumbar most common; enlarged dense trabeculae, especially around the contours, give a *"picture frame" appearance*. If it is more uniformly dense, the increased body size helps differentiate it from metastasis or other etiology of an ivory vertebra. May develop compression fractures.
2. *Cranium (65% of patients)*: Lytic phase, termed *osteoporosis circumscripta*, may start in either the frontal or occipital areas. Later, focal radiodense areas develop (*"cotton wool"* appearance). *Basilar invagination in one third.*
3. *Pelvis (70% of patients)*: Initial finding is *cortical thickening* along the *ileopectineal line*; if unilateral involvement, the asymmetrical size increase becomes obvious. The softened bone develops a protrusio acetabuli deformity.
4. *Tubular bones (35% of patients)*: *Lysis always begins in subarticular region*, advancing with a "flame" or "blade of grass" border. *Exception to this is the tibia, where the initial lysis may be diaphyseal.* Bone enlarges and bows.

G. Musculoskeletal complications:

1. *Stress fracture*: Horizontal lines on the convex aspect of bone.
2. Osteoporosis if immobilized.
3. Neoplasm:
   a. Involved bone may degenerate to *sarcoma* (osteosarcoma more commonly than fibrosarcoma); a rare occurrence (probably

less than 1% overall) but may be seen in
5% to 10% of patients with widespread
disease.
  b. *Giant cell tumors* (usually benign) involving
     the facial bones and cranium.
4. Osteoarthritis due to mechanical
   abnormalities.

XIII. DRUG- AND ENVIRONMENTALLY INDUCED ABNORMALITIES OF BONE

A. Drugs
  1. Coumadin: Embryopathy with stippled
     epiphyses.
  2. Heparin: Osteoporosis (dose-related).
  3. Dilantin and phenobarbital: Osteomalacia.
  4. Steroids: Osteoporosis, AVN.
  5. Lead: Widened metaphyses with dense
     bands, which may be multiple from several
     episodes of lead poisoning. A single dense
     metaphyseal line may be physiologic, but if
     one is present at the fibular metaphysis as
     well as other metaphyses around the knee, it
     is more likely to represent lead poisoning
     than a physiologic finding.
  6. Bismuth: Same as lead.
  7. Vitamin A poisoning: Abnormalities appear
     only after 6 months of age. Periostitis of long
     bones (painful) is the first abnormality. An
     older child may have a growth disturbance
     and coned epiphyses.
  8. Vitamin D poisoning: Increased density and
     periostitis; prominent soft tissue calcifications.
     Osteoporosis also may be present.
  9. Fluorosis: Increased density, periostitis,
     ossification of ligaments (especially pelvic);
     stress fractures.
  10. Alcohol: Fetal alcohol syndrome (delayed
      development, among other findings); in
      adult, osteoporosis and AVN.
  11. Prostaglandins: Periostitis in infants.

XIV. Serum Laboratory Values in Metabolic Bone Disease

	Calcium*	Phosphate*	Alkaline Phosphatase*
Osteoporosis	nl	nl	nl
Osteomalacia/rickets	↓	↓	↑
Renal osteodystrophy (secondary HPTH)	nl to ↓	↑	↑
Hypophosphatasia	nl	nl	↓
Paget's disease	nl	nl	↑
Hypoparathyroidism	↓	↑	nl
Pseudohypoparathyroidism	↓	↑	nl
Pseudo-pseudohypoparathyroidism	nl	nl	nl

*nl = normal; ↑ = increased; ↓ = decreased.

B. Environmental causes
  1. Polyvinylchloride: Acro-osteolysis.
  2. Burns: Contractures, acro-osteolysis, heterotopic ossification.
  3. Frostbite: Acro-osteolysis (dense or resorbed distal phalangeal epiphyses in children); thumb often is spared.

REFERENCES

1. Riggs B, Melton L: Involutional osteoporosis. *N Engl J Med* 1986; 314:1676–1684.

BIBLIOGRAPHY

Resnick D, Niwayama G: *Diagnosis of Bone and Joint Disorders*, ed 2. Philadelphia, W. B. Saunders, 1988.
Jacobson H, Edeiken J (eds): *Syllabus—Metabolic and Endocrine Disorders Affecting the Skeleton*. Radiological Society of North America 67th Scientific Assembly, November 1981.

5 | Congenital Anomalies

The multitude of congenital syndromes is confusing. Furthermore, there is overlap in radiographic abnormalities among the syndromes. The more common congenital abnormalities are discussed briefly in this section; a practical approach to classifying dwarfs is also presented. An excellent reference for wider gamuts and less common syndromes is Taybi and Lachman: *Radiology of Syndromes, Metabolic Disorders, and Skeletal Dysplasias*, ed 3 (Chicago, Year Book Medical Publishers, 1989). In this book, cross indexes reference the syndromes and give gamuts according to the nature of radiographic abnormalities as well as their sites.

In addition, many apparent abnormalities may be normal variants or, in children, a transitional state seen only at a certain age. The most complete reference for normal variants is Keats: *An Atlas of Normal Roentgen Variants That May Simulate Disease*, ed 4 (Chicago, Year Book Medical Publishers, 1988). The best reference for different appearances of the skeleton through different periods of skeletal maturation is Keats and Smith: *An Atlas of Normal Developmental Roentgen Anatomy* (Chicago, Year Book Medical Publishers, 1988).

I. SCOLIOSIS

> **Key Concepts:** Must evaluate for intrinsic vertebral abnormalities, tumor, and neurofibromatosis. The latter disease may cause the spine to collapse rapidly, producing paraplegia. Posteroanterior (PA) views expose patients' breasts to less ionizing radiation than anteroposterior (AP) views.

A. Definition: Lateral deviation and rotation of the spine, often associated with thoracic hypokyphosis. Severe disease distorts the chest wall enough to restrict pulmonary and cardiovascular function, requiring correction. The cosmetic deformity may also be serious.

B. Etiology: Must be determined in each case of scoliosis since prognosis and treatment are highly dependent on etiology.

  1. Idiopathic:
    a. By far the most common (85%).
    b. Thought to be related to defects in proprioception and vibratory sense.
    c. Girls are more often affected than boys (7:1) and their disease is more likely to progress and require treatment.
    d. Patterns may be thoracic (usually convex to the right), thoracolumbar, lumbar, or double major.
    e. Infantile idiopathic is uncommon, occurs more often in males, is usually thoracic and convex to the left, and is difficult to treat. Congenital forms must be ruled out.
    f. Juvenile (ages 3 to 10 years) and adolescent idiopathic scoliosis are much more common. Curves that are greater than 50 degrees or that show progression usually require surgical treatment. The curves are generally flexible (demonstrated

by lateral bending films), and L_5 spondylolysis may be associated.

2. Congenital:
    a. Secondary to vertebral anomalies—hemivertebrae (failure of formation) or bars (failure of segmentation) causing unbalanced growth.
    b. May have associated tethered cord or diastematomyelia; magnetic resonance imaging (MRI) makes the diagnosis.
    c. Bracing is ineffectual, and progressing curves must be fused early.

3. Neuromuscular:
    a. Etiology may be spasticity, paralysis, arthrogryposis, or muscular dystrophy.
    b. The curves are long, single curves.

4. Diseases of collagen synthesis:
    a. Marfan's syndrome.
    b. Ehlers-Danlos syndrome.
    c. Homocystinuria.

5. Neurofibromatosis:
    a. Dysplastic vertebral bodies form a short, angular, usually high thoracic curve.
    b. Associated ribbon-shaped ribs and posterior vertebral body scalloping help make the diagnosis.
    c. These curves must be monitored carefully since they can collapse rapidly and produce paralysis.

6. Trauma: May require instrumentation if there is a burst fracture, especially in the lumbar spine.

7. Tumors:
    a. Osteoid osteoma is the most common tumor that causes scoliosis.
    b. The lesion may be lucent but there is surrounding sclerosis. It is usually located in posterior elements.
    c. The scoliosis is long, with concavity on the side of the lesion. There is no rotational component.

        d. Bone scan may help locate an occult
           lesion.
   8. Radiation therapy: If entire vertebral body of
      a skeletally immature patient is not included
      in a radiation port, growth in the irradiated
      portion stops, with resultant scoliosis.

C. Filming:
   1. The initial film should be good quality AP
      erect films, in order to evaluate for intrinsic
      vertebral abnormalities.
   2. Subsequent films used only to evaluate
      progression should be high kVp and must be
      taken PA rather than AP, using gradient
      intensifying screens or filters. Gonad and
      breast shields should be utilized if possible.
      Collimation is essential. This approach
      reduces the dose to breast tissue to one sixth
      to one tenth that of AP films. The breast is
      the most radiation-sensitive tissue in
      adolescent females and should be protected.

D. Treatment:
   1. Bracing.
   2. Electrical stimulation.
   3. Posterior fusion: Rodding must be
      accompanied by sufficient graft for a fusion
      mass to form.
      a. Harrington rods: ratcheted portion is
         cephalad. Most common site of failure is
         at the junction of the ratcheted and
         smooth portion. Superior and inferior
         shoes are different in size and shape; they
         also must be watched for detachment from
         the pedicle or laminae.
      b. Luque rods, Us, or squares are another
         posterior system that is secured either by
         sublaminar wiring (extremely secure but
         risks tearing dura during removal) or
         Wisconsin segmental spinal instruments
         (WSSI) wiring (through the thick portion
         of the spinous process, secured by
         buttons; almost as secure as sublaminar

wiring and safer to place and remove).
c. Anterior rodding (Dwyer, Zielke, Dunn) rarely used for nontraumatic scoliosis.
d. Cotrel-Dubousset rodding: A complicated posterior rod system that reduces the scoliosis by correcting the rotatory deformity. This protects the thoracic kyphosis and is the only system that also corrects the rib hump deformity of scoliosis.

II. ARTHROGRYPOSIS MULTIPLEX CONGENITA

Key Concepts: Long scoliosis, multiple dislocations and extremity deformities; atrophic soft tissues with webbed joints and "dense" capsules.

A. Etiology: The cause is disputed. Decrease in size and number of anterior horn cells; fibro-fatty changes in musculature are seen histologically.
B. Clinical features:
 1. Apparent in utero or at birth.
 2. Muscle wasting.
 3. Symmetric contractures of appendicular joints.
 4. Absence of normal skin creases.
 5. Webbing of skin when joints are fixed in flexion.
 6. Thickened articular capsules.
 7. Intelligence usually normal.
 8. Clubfoot, congenital dislocation of the hip (CDH), rudimentary or absent patellae, club hand, flexed fingers, adducted thumbs.
C. Radiographic findings:
 1. Atrophic musculature with relatively increased density of joint capsules.
 2. Typical deformities:
  a. Fixed flexion deformities.

    b. Scoliosis (long, neuromuscular type).
    c. Hip subluxation/dislocation.
    d. Rudimentary or absent patellae.
    e. Clubfoot (or may have valgus deformities).
    f. Elbow dislocations.
    g. Club hand in ulnar deviation.
    h. Carpal fusion.
    i. Adducted thumb.
  3. Osteoporosis.
  4. Frequent fractures although bone is normal aside from being osteoporotic.
D. Distribution:
  1. Both upper and lower extremity disease: 50%.
  2. Upper extremity disease only: 40%.
  3. Lower extremity disease only: 10%.

III. NEUROFIBROMATOSIS

> **Key Concepts:** A bone dysplasia with multiple manifestations; the kyphoscoliosis is the most significant since it may progress rapidly to collapse.

A. Skeletal involvement in 80% (90% if macrocranium is included).
B. Kyphoscoliosis in 50%:
  1. T_{3-7} most common.
  2. Short segment and angular.
  3. Need not be present at birth.
  4. Attributed to a primary mesodermal dysplasia.
  5. May be rapidly progressive, causing paraplegia.
C. Posterocentral vertebral body scalloping with erosion of pedicles.
  1. May be progressive.
  2. Usually due to dural ectasia, with or without the presence of neurofibromas.
D. Loosely attached periosteum is easily stripped causing large calcified subperiosteal hemorrhage

or poor callus response and pseudarthrosis.

E. Bowing and pseudarthrosis of the tibia, most commonly at the junction of the middle and distal thirds; an abnormally formed, deficient, gracile fibula with pencil-pointing of the segments is often seen, with or without a tibial pseudarthrosis.

F. Irregular, notched, scalloped, twisted, ribbonlike ribs are usually due to the primary bone dysplasia, but occasionally there is a neighboring intercostal neurofibroma.

G. Localized gigantism (occasionally dwarfism).

H. "Cystic" lesions are controversial.
  1. May relate to deossification of bone, reaction to subperiosteal hemorrhage, or local erosion rather than replacement by tumor.
  2. Biopsy of some large, expanded lesions has proven them to be nonossifying fibromas.

I. Small exostoses may arise adjacent to a soft tissue neurofibroma.

J. Local erosion due to a soft tissue neurofibroma.

K. Skull.
  1. Deficient or absent sphenoid wing.
  2. Hypoplasia of posterosuperior orbital wall with herniation into posterior orbit and exopthalmos.
  3. Cranial defects, most commonly at the left lambdoid suture, sometimes accompanied by an ipsilateral hypoplastic mastoid, are due to underlying mesodermal dysplasia rather than neurofibromatous tissue.
  4. Enlargement of cranial foramina, especially optic and internal auditory canals, due to neuromas or gliomas.
  5. Macrocranium in 75%.

IV. MARFAN'S SYNDROME

> ***Key Concepts:*** Connective tissue abnormality with ocular (bilateral ectopic lenses), cardiovascular (ascending aortic dissection, valvular insufficiency), and musculoskeletal (arachnodactyly, kyphoscoliosis, posterior vertebral body scalloping) abnormalities.

A. Definition: Familial (usually autosomal-dominant) *connective tissue disorder* in which the primary defect is unknown. Patients are tall and thin, with *disproportionately long extremities, especially distally (arachnodactyly)*.
B. Epidemiology: No race or sex preponderance.
C. Nonmusculoskeletal features:
   1. Ocular: Ectopic lenses.
   2. Cardiovascular: Cystic medial necrosis of the ascending aorta or pulmonary artery predisposes to dissection and rupture. Aortic and mitral valve insufficiency also occur.
D. Musculoskeletal features:
   1. *Arachnodactyly.*
   2. *Kyphoscoliosis* in a pattern similar to that of idiopathic scoliosis but occurring earlier.
   3. *Posterior scalloping of vertebral bodies* (secondary to dural ectasia).
   4. *Spondylolysis*, especially of L_5.
   5. *Pectus excavatum.*
   6. *Normal bone density*: This feature is important in differentiating Marfan's syndrome from homocystinuria (arachnodactyly with osteoporosis).
   7. Deformities may occur secondary to hypermobility (pes planus, genu recurvatum, patella alta), as can dislocations.
   8. Hypermobility may lead to premature osteoarthritis, but otherwise the joints are normal.

V. HOMOCYSTINURIA

> ***Key Concepts:*** Familial connective tissue disorder with ocular (lens dislocation) and skeletal abnormalities resembling Marfan's syndrome (disproportionately long extremities, scoliosis, joint laxity). Major differences are the frequent thrombotic episodes and osteoporosis found in homocystinuria.

A. Definition: A *familial* (autosomal-recessive) disease resulting from deficiency of the enzyme cystathionine synthetase. There is an associated *defect in collagen* synthesis, which affects multiple organ systems.
B. Epidemiology: More common in patients of Northern European extraction.
C. Nonmusculoskeletal features:
   1. *Mental retardation* and seizures.
   2. Frequent, often life-threatening, *thrombotic episodes* (venous and arterial) secondary to abnormal platelet aggregation.
   3. *Lens dislocation,* often bilateral.
   4. Cystic medial necrosis in all elastic arteries, but dissections are rare (unlike Marfan's).
D. Musculoskeletal features:
   1. *Arachnodactyly* (as in Marfan's).
   2. *Scoliosis.*
   3. AP diameter of vertebral bodies may be increased, with *posterior scalloping.*
   4. Pectus excavatum.
   5. *Osteoporosis* is an important feature that is constant in homocystinuria but not seen in Marfan's; end-plate biconcavity and *vertebral compression fractures are common.*
   6. Joint laxity occurs, as in Marfan's, but *flexion contractures* are more common.

VI. EHLERS-DANLOS SYNDROME

> *Key Concepts:* Familial connective tissue disorder with many features similar to those of Marfan's syndrome (arachnodactyly, kyphoscoliosis, posterior vertebral scalloping, spondylolysis). Subcutaneous calcifications and the history of skin hyperelasticity help make the correct diagnosis.

A. Definition: Familial (usually autosomal-dominant) spectrum of connective tissue diseases characterized by hyperelasticity of the skin, laxity of the joints, and a bleeding disorder, affecting multiple systems; there is an associated defect in collagen synthesis.

B. Epidemiology: Usually patients are Caucasians of European origin; males are affected predominantly.

C. Nonmusculoskeletal features:
   1. *Thin, hyperelastic skin.*
   2. *Fragile vessel walls predispose patient to bleeding and dissecting aneurysms.*

D. Musculoskeletal features:
   1. No primary osseous abnormality.
   2. *Ligamentous laxity allows subclinical trauma to joints, as well as dislocations. Early osteoarthritis results.*
   3. *Kyphoscoliosis and other deformities* (genu recurvatum and pes planus are the most common).
   4. *Posterior vertebral scalloping.*
   5. *Arachnodactyly.*
   6. *Spondylolysis.*
   7. *Subcutaneous calcification,* most often *in forearms and shins,* result from fat necrosis and hematomas. Calcifications look like phleboliths.

VII. OSTEOGENESIS IMPERFECTA

> **Key Concepts:** Spectrum of connective tissue diseases with both congenital and late forms. Blue sclerae, osteoporosis, wormian bones, and multiple fractures are the most common abnormalities.

A. Definition: A congenital connective tissue abnormality (both osteoid and collagen are abnormal) which is manifested in a wide clinical spectrum. This spectrum has been divided into three main groups but there is overlap among them.

B. Epidemiology: No race or sex predilection.

C. Features common to all three groups:
1. *Blue sclerae.*
2. Severe *osteoporosis* and resultant multiple *fractures.*
3. *Elevated alkaline phosphatase* secondary to multiple fractures.
4. *Exuberant callus formation* and fracture healing (though often with deformity).
5. *Wormian bones* in the skull.
6. Dental abnormalities.
7. Thin skin.
8. Joint laxity.
9. Otosclerosis.
10. Platelet abnormalities and vascular structural abnormalities resulting in bleeds.
11. Platybasia, wedge deformities of the vertebral bodies, and kyphoscoliosis.

D. Three major groups:
1. *Osteogenesis imperfecta congenita.*
   a. *Autosomal-recessive disorder.*
   b. *Long bones are short, thick, and bowed* (resembling dwarfism).
   c. *Multiple fractures at birth* (usually acquired in utero).

      d. Large skull.
      e. Infants are stillborn or die very early.
      f. Rare survivors may develop cystic
         metaphyses.
  2. *Osteogenesis imperfecta tarda 1.*
      a. *Autosomal-dominant* disorder.
      b. *Long bones are thin and of normal length.*
      c. *May have fractures at birth but not prenatally.*
      d. Fracture rate decreases at puberty.
  3. *Osteogenesis imperfecta tarda 2*: Similar in
     appearance to osteogenesis imperfecta tarda 1
     but with *fewer fractures* and *later onset*.

VIII. SCLEROSING DYSPLASIAS

> **Key Concepts:** Spectrum of diseases with various patterns of increased osseous density. Some are asymptomatic; others cause brittle bones that develop pathologic fractures.

A. Definition: A *spectrum of diseases* resulting from a
   failure of osteoclasts to resorb bone during
   remodeling. This failure may occur at sites of
   either endochondral or intramembranous
   ossification.
B. Individual diseases in the spectrum:
  1. *Osteopetrosis* (Albers-Schonberg disease).
      a. *Sclerotic, fragile long bones with frequent
         transverse fractures.*
      b. Transverse dense metaphyseal bands are
         common.
      c. May have *bone-within-a-bone* appearance.
      d. Metaphyseal flaring is secondary to
         abnormal remodeling.
      e. Infantile autosomal-recessive form is
         severe and often fatal; adult form, which
         may be recessive or dominant, is more
         benign.
      f. Patient is anemic and subject to infection.
      g. Sclerosis and lack of remodeling of

surrounding bone cause cranial nerve damage (blindness and deafness).

2. *Pyknodysostosis.*
   a. Similar to osteopetrosis with *sclerotic long bones that sustain transverse pathologic fractures.*
   b. Additional features are *wormian bones, hypoplastic angle of the jaw, and acro-osteolysis.*
3. *Osteopoikilosis: Multiple bone islands located in the epiphysis and metaphysis.*
4. Osteopathia striata (Voorhoeve's disease): Linear striations in the metaphyses.
5. Progressive diaphyseal dysplasia (Camurate-Engelmann disease):
   a. Symmetrically thickened cortex with sparing of the metaphyses and epiphyses.
   b. Both the endosteum and periosteum are involved, narrowing the medullary cavity.
6. Ribbing disease: Asymmetric, painful cortical thickening.
7. Melorheostosis.
   a. A wavy hyperostosis ("dripping candle wax") that is endosteal as well as periosteal, involving only one side of the bone.
   b. Lower extremities more commonly affected than other sites.
   c. Unilateral predisposition, tendency to involve multiple bones in the same ray; tarsals may be involved as well.

IX. CONGENITAL DISLOCATED HIP (CDH)

Key Concepts: Lateral and superior migration of the femoral head, often associated with acetabular dysplasia and hypoplasia of the femoral head ossification center.

A. Definition: Perinatal relaxation of the capsule of the hip joint allows dislocation of the hip. Acetabular deficiency, femoral head and neck deformity, and contractures of periarticular muscles are secondary features.
B. Epidemiology.
   1. *Caucasians* have the highest incidence.
   2. Incidence greater in *females* than males 5:1.
   3. Breech delivery and oligohydramnios are predisposing factors.
   4. Very common in neonatal period, but 75% to 95% of newborns with clinical signs of CDH revert to normal after a few weeks.
   5. Incidence after neonatal period 1:1000.
C. Radiographic signs:
   1. *Films prior to 6 weeks of age have a high false-negative rate.* Early films may, however, be useful in ruling out other etiologies of leg shortening (such as proximal focal femoral deficiency).
   2. *Primary radiographic signs[1] of CDH* (Fig 5–1):
      a. Draw *Hilgenreiner's line* (through triradiate cartilages), then *Perkin's line* (from the anterior inferior iliac spine perpendicular to Hilgenreiner's line); *normally the femoral head ossification center is located in the inner lower quadrant formed by these lines.*
      b. Measure horizontal distance from teardrop to metaphysis and compare to normal side; a difference of 2 mm or more indicates *lateral subluxation* of the hip.
      c. Measure vertical distance from Hilgenreiner's line to metaphysis; a difference of 2 mm or more indicates superior subluxation of the hip. *If the hip is displaced both laterally and superiorly, it is considered dislocated.*
      d. *Shenton's line* (following the curve from the obturator foramen extended to the femoral neck) may be helpful.

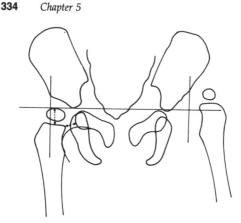

FIG 5–1.
Left CDH; normal right hip. Hilgenreiner's line (Y line through triradiate cartilages) serves as a reference. The left hip shows both lateral and superior displacement (as measured from the teardrop and Hilgenreiner's line, respectively) and is, therefore, dislocated. Shenton's line (shown on the normal right side outlining the obturator foramen and femoral neck) would be distorted on the abnormal left side.

2. *Secondary radiographic signs of CDH:*
   a. *Acetabular dysplasia:* Present when the angle described by Hilgenreiner's line and the acetabular roof is greater than 30 degrees. This "shallow acetabulum" is rarely present initially but develops because of inadequate stress by the malpositioned femoral head; often reconstitutes normally after reduction of the hip dislocation.
   b. Excessive *anteversion* of the femoral head. The 30 degrees' anteversion of the head at birth normally decreases to 10 degrees by

the time of skeletal maturity. This process often fails to occur in CDH.

    c. Delayed ossification of the proximal femoral epiphysis is often present but nonspecific. Normal femoral heads may show asymmetric ossification.

D. Complications of CDH:
   1. Avascular necrosis (AVN) secondary to manipulation; tenotomy and traction may reduce the risk.
   2. Development of pseudoacetabulae in the iliac wings and associated mechanical problems.
   3. Secondary osteoarthritis occurs surprisingly late, generally in the fifth or sixth decade.

E. *Ultrasound*, in experienced hands, is extremely accurate.[2]

F. *Arthrography* is employed in patients with persistent and unexplained CDH, primarily to demonstrate sources of mechanical obstruction to a successful reduction.
   1. *Inverted limbus "thorn"* (superior acetabular labrum).
   2. *Pulvinar* (intracapsular soft tissue).
   3. *Elongated ligamentum teres* and *hourglass constriction* of the joint capsule.
   4. Tight *iliopsoas* muscle.

G. Computed tomography (CT) or MRI is occasionally useful in further defining abnormal bone relationships, especially deficiency in the posterior rim of the acetabulum.

H. Treatment of CDH.
   1. Closed reduction.
   2. Skeletal traction.
   3. Adductor tenotomy and iliopsoas release.
   4. Open reduction for mechanical obstruction.
   5. Varus derotational osteotomy of the femoral subtrochanteric region.
   6. Pemberton acetabuloplasty: An osteotomy extending from the anterior inferior iliac spine to the triradiate cartilage, using the latter as a

hinge. An opening wedge is placed. This reconstructive osteotomy for the skeletally immature patient affords coverage of the femoral head and allows congruent growth of the femoral head and acetabulum.

7. Salter opening wedge osteotomy: A reconstructive osteotomy that is less difficult to perform than the Pemberton. The osteotomy extends from the anterior inferior iliac spine straight across to the sacrosciatic notch. An opening wedge is placed, using the symphysis pubis as a hinge. As with the Pemberton operation, the acetabular roof is shifted anterolaterally to cover the femoral head.

8. Triple innominate (triple Steele) osteotomy: A reconstructive procedure for the skeletally mature patient. Osteotomies are performed across the iliac neck, pubis, and ischium, leaving a free-floating acetabulum, which is rotated so that femoral head coverage is attained and secured with hardware at the iliac neck osteotomy site.

9. Chiari medial displacement: This is not a reconstructive, but a salvage procedure for older patients. An intra-articular osteotomy is extended across the superior acetabulum. The inferior portion of the osteotomy as well as the femoral head are displaced medially, thus attaining the goal of femoral head coverage.

X. PROXIMAL FEMORAL FOCAL DEFICIENCY

A. Definition: A spectrum of abnormalities involving agenesis of all or part of the proximal femur. The entity usually is an isolated abnormality and unilateral.

B. Radiographic abnormalities.
   1. Short, displaced femur.
   2. Normal distal femur.
   3. Proximal femoral abnormality ranges from

complete absence to a large gap between the diaphysis and capital femoral epiphysis, to pseudarthrosis at the femoral neck, to a varus deformity. Imaging procedures seek to classify the abnormality early so that prognosis and treatment may be established. In younger patients, where apparently absent structures may in fact be cartilaginous, arthrography or MR may demonstrate nonossified structures.
4. Major differential diagnosis is infantile coxa vara.

XI. INFANTILE COXA VARA

A. Definition: Development of coxa vara (neck-to-shaft angle of less than 120 degrees) in infancy, often related to walking. Etiology is unknown, and the abnormality may be progressive; 25% are bilateral. Leg-length discrepancy and limp develop.
B. Differential diagnosis:
1. Proximal femoral focal deficiency, mild.
2. Renal osteodystrophy and rickets.
3. Coxa vara in cleidocranial dysostosis.
4. Slipped capital femoral epiphysis. At so young an age, it is more likely due to infection. An idiopathic variety is seen in adolescents.

XII. PRIMARY PROTRUSIO OF THE ACETABULUM

A. Definition: Protrusio acetabuli without any recognized etiology and no other radiographic abnormality; perhaps due to abnormal acetabular remodeling; usually bilateral and more common in women; often familial; secondary osteoarthritis is common.
B. Differential diagnosis.
1. Arthritides with axial migration—rheumatoid arthritis (RA), rheumatoid variants, 20% of

degenerative joint disease (DJD).
2. Diseases that soften bone, leading to protrusio, such as osteomalacia or Paget's disease of bone.

XIII. PSEUDARTHROSIS

A. Definition: Interruption in the diaphysis of a long bone, usually with tapering ends and mobility at the site of pseudoarthrosis.
B. Pseudarthrosis is most common in the tibia, followed by the fibula and clavicle; may be congenital or may develop in infancy.
C. May be an isolated phenomenon or related to neurofibromatosis or fibrous dysplasia.
D. If it develops in an infant, there is anterior bowing first, then pathologic fracture, then tapering of the bones at the fracture site.

XIV. CONGENITAL FOOT ANOMALIES

Key Concepts: Most foot anomalies can be described using the following parameters: hindfoot equinus or calcaneus, hindfoot varus or valgus, forefoot varus or valgus. Clubfoot consists of hindfoot equinus, hindfoot varus, and forefoot varus. Congenital vertical talus deformity consists of hindfoot equinus, hindfoot valgus, forefoot valgus, and talonavicular dislocation. All of these parameters should be evaluated *only on weight-bearing films.*

A. Terminology:[3, 4]
   1. *Cavus*: Raised longitudinal arch of the foot (literally, *hollow*).
   2. *Planus*: Flattened longitudinal arch of the foot.
   3. *Equinus*: Fixed plantar flexion of the hindfoot.
   4. *Calcaneus*: Fixed dorsiflexion of the hindfoot.
   5. *Varus (adductus)*: Inverted (hindfoot or forefoot).

6. *Valgus (abductus)*: Everted (hindfoot or forefoot).
7. *Supination of the forefoot*: Inward rotation of forefoot on hindfoot (sole faces inward).
8. *Pronation of the forefoot*: Excessive outward rotation of forefoot on hindfoot.

B. Most valuable measurements are *all made on weight-bearing films.*
1. *Hindfoot equinus or calcaneus:*
   a. Normal on the lateral film: the calcaneus is normally dorsiflexed. The angle between the longitudinal axis of the tibia and the calcaneus (measured along the base) ranges between 60 and 90 degrees (Fig 5–2,A).
   b. *Hindfoot equinus*: Plantar flexion of the calcaneus such that the calcaneotibial angle is greater than 90 degrees (Fig 5–2,B); seen in clubfoot and congenital vertical talus.
   c. *Hindfoot calcaneus*: Excessively dorsiflexed calcaneus such that the calcaneotibial angle is less than 60 degrees (Fig 5–2,C); seen in cavus and spastic deformities.
2. *Hindfoot varus or valgus*: Talus is the point of reference since it is assumed to be fixed relative to the lower leg; calcaneus rotates around it.
   a. *Normal.*
      (1) *Lateral*: Talocalcaneal (TC or Kite's) angle 25 to 45 degrees (50 degrees in newborns; Fig 5–3,A).
      (2) *AP*: TC angle 15 to 40 degrees (30 to 50 degrees in newborns). The midtalar line passes through or slightly medial to the base of the first metatarsal (MT). The midcalcaneal line passes through the base of the fourth MT (Fig 5–4,A).
   b. *Hindfoot varus*: With fixed talus, the calcaneus can adduct. Hindfoot varus is

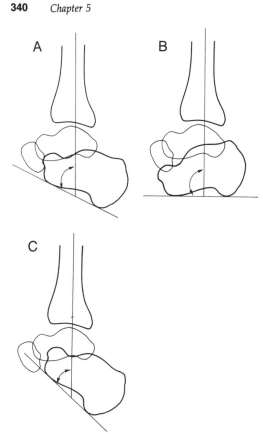

FIG 5–2.
A, normal hindfoot with the tibiocalcaneal angle measuring between 60 and 90 degrees. B, hindfoot equinus (angle greater than 90 degrees). C, hindfoot calcaneus (angle less than 60 degrees).

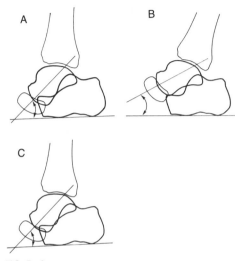

FIG 5–3.
Lateral view: **A,** normal hindfoot with the lateral talo-calcaneal (TC) angle measuring 25 to 50 degrees. **B,** hindfoot varus with a decreased lateral TC angle. **C,** hindfoot valgus with an increased lateral TC angle.

seen in clubfoot and some paralytic deformities.

(1) *Lateral: Decreased TC angle* (less than 25 degrees), with the talus and calcaneus, approaching parallelism (Fig 5–3,B).

(2) *AP: Decreased TC angle* (less than 15 degrees) (Fig 5–4,B). The talus points lateral to the first MT because the entire foot swings medially.

c. *Hindfoot valgus*: With fixed talus, the calcaneus can abduct. Hindfoot valgus is seen in congenital vertical talus, flexible

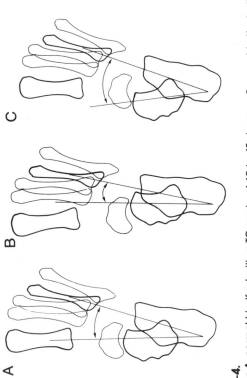

FIG 5–4.
AP view: **A,** normal hindfoot with a TC angle of 15 to 40 degrees. **B,** varus hindfoot with a decreased TC angle. **C,** valgus hindfoot with an increased TC angle.

flatfoot deformity, and neurologic
deformities.
   (1) *Lateral: Increased TC angle* (greater than
       45 degrees, or, in newborns, 50
       degrees) (Fig 5–3,C).
   (2) *AP*: Increased TC angle (Fig 5–4,C).
       The talus points medial to the first MT
       because the calcaneus (and, therefore,
       the entire foot) swings laterally.
3. *Forefoot varus or valgus.*
   a. *Normal.*
      (1) *AP*: MTs converge proximally with
          slight overlap at the bases (Fig 5–5,A).
      (2) *Lateral*: Fifth MT is in most plantar
          position, with other MTs superimposed
          (Fig 5–6,A).
   b. *Forefoot varus* (inverted, often supinated).
      Forefoot varus is seen in clubfoot
      deformities and spastic deformities.
      (1) *AP*: Forefoot is narrowed, with
          increased convergence at bases (Fig 5–
          5,B).
      (2) *Lateral*: Ladderlike arrangement, with
          first MT most dorsal and fifth MT most
          plantar (Fig 5–6,B).
   c. *Forefoot valgus* (everted, often pronated).
      Forefoot valgus is seen in congenital
      vertical talus, flexible flatfoot, and spastic
      deformities.
      (1) AP: Forefoot is broadened with
          decreased overlap at bases (Fig 5–5,C).
      (2) Lateral: May see ladderlike
          arrangement, but first MT is in most
          plantar position (Fig 5–6,C).
C. Common foot deformities.
   1. Clubfoot (talipes equinovarus).
      a. Incidence: One in 1000 births.
      b. Ratio of males to females affected: 2–3:1.
      c. Etiology unclear. Possible contributing
         factors are:

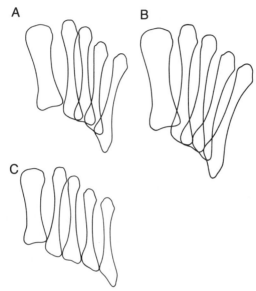

FIG 5–5.
AP view: **A,** normal forefoot with mild MT convergence at bases. **B,** forefoot varus with increased MT convergence. **C,** forefoot valgus with decreased convergence.

        (1) Defective connective tissue with ligamentous laxity.
        (2) Muscle imbalance.
        (3) Intrauterine position deformity.
        (4) Persistence of an early normal fetal relationship.
    d. Radiographic findings (Fig 5–7):

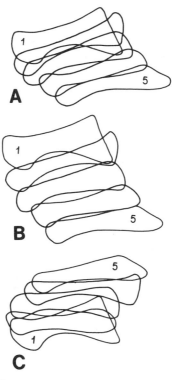

FIG 5–6.
Lateral view: **A,** normal forefoot with MT superimposition (but fifth in plantar position). **B,** forefoot varus with ladder configuration of MTs and 5th in plantar position. **C,** forefoot valgus with MT superimposition (but first MT in plantar position).

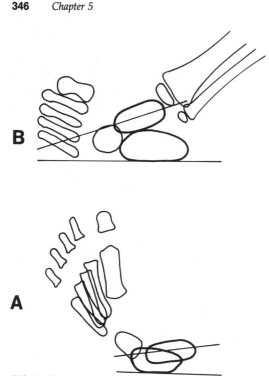

FIG 5–7.
Clubfoot deformity: hindfoot equinus, hindfoot varus,
forefoot varus. **A,** AP. **B,** lateral.

   (1) Hindfoot equinus.
   (2) Hindfoot varus.
   (3) Forefoot varus (more or less,
       depending on severity of deformity).
 2. Congenital vertical talus (rocker-bottom foot).
    The talus is in extreme plantar flexion with
    dorsal dislocation of the navicular, locking the

talus into plantar flexion.
   a. Rigid flat foot.
   b. May occur in isolation or as part of a
      variety of syndromes, frequently
      associated with myelomeningocele.
   c. Radiographic findings (Fig 5–8):
      (1) *Lateral*:
          a Abnormal talus with *dislocated
             navicular.*
          b Valgus hindfoot.
          c Equinus.
          d Forefoot dorsiflexed and valgus.
      (2) *AP*:
          a Severe hindfoot valgus.
          b Forefoot normal to valgus.
3. Flexible flatfoot deformity (pes planovalgus).
   a. Affects 4% of population.
   b. Hereditary influence.
   c. Etiology may be lax ligaments that allow
      the calcaneus to rotate into a valgus
      position. The talus becomes more vertical
      but the navicular follows it.
   d. Radiographic findings: (Note: It is flexible:
      non–weight-bearing films are normal.)
      (1) AP:
          a Hindfoot valgus.
          b Forefoot valgus with flattened
             midtarsal arch.
      (2) Lateral:
          a Hindfoot valgus.
          b May superficially resemble
             congenital vertical talus, but the
             talonavicular joint is not dislocated
             and there is no equinus.
4. Pes cavus: High arched foot (calcaneus) with
   compensatory plantar flexion of the forefoot.
   a. Etiology:
      (1) Upper motor neuron lesions
          (Friedreich's ataxia).
      (2) Lower motor neuron lesions

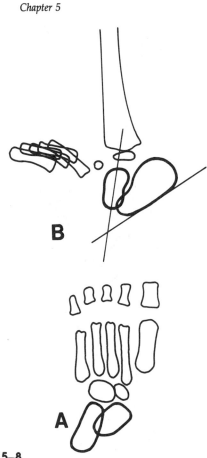

FIG 5–8.
Congenital vertical talus: hindfoot equinus, hindfoot valgus with dislocated talonavicular joint, forefoot valgus. A, AP. B, lateral.

(poliomyelitis).
(3) Vascular ischemia as in Volkmann's contracture.
(4) Muscular dystrophy of the peroneal type (Charcot-Marie-Tooth disease).
5. Metatarsus adductus.
  a. Most common structural abnormality of the foot seen in infancy.
  b. Ten times as common as clubfoot.
  c. Usually bilateral.
  d. Etiology unknown.
  e. More common in females.
  f. Radiographic findings:
    (1) *Forefoot adducted and in varus.*
    (2) *Hindfoot normal* to moderate valgus.
    (3) No equinus (dorsiflexion normal).

XV. TARSAL COALITION (PERONEAL SPASTIC FLATFOOT)

> **Key Concepts:** May be the etiology of painful flatfoot in young males. Secondary changes of sclerosis and beaking are seen on lateral film. Coalition seen on oblique (calcaneonavicular) or Harris (talocalcaneal) views; CT is extremely useful.

A. Epidemiology:
  1. Occurs in 1% of population.
  2. May be familial, but with great variability.
  3. May be congenital (vast majority) or secondary to infection, trauma, arthritis disorders, or surgery.
  4. May be a part of various syndromes: hereditary symphalangism, Aperts' acrocephalosyndactyly, hand-foot-uterus.
  5. Congenital probably due to a failure of segmentation in the fetus. (Coalitions are seen in fetuses.)

6. Calcaneonavicular more common than talocalcaneal, more common than talonavicular, more common than calcaneocuboid.
7. Males more commonly affected.
8. Bilateral 25%.

B. Signs and symptoms:
1. Symptoms generally first occur in the second or third decade.
2. Physical examination findings: Limited subtalar motion, pes planus, shortening with persistent or intermittent spasm of the peroneal muscles.

C. Radiographic findings:
1. Coalition may be fibrous, cartilaginous, or osseous, so may not demonstrate bony bridging in all cases. In the absence of a bony bridge, close approximation of the bones with cortical irregularity or sclerosis suggests fibrous or cartilaginous bridging.
2. Calcaneonavicular coalition.
   a. Symptoms are less severe than in a talocalcaneal coalition.
   b. Secondary radiographic signs are less marked.
   c. Talar "beaking" uncommon.
   d. Best seen on 45-degree medial oblique (Fig 5–9).
3. Talocalcaneal coalition.
   a. Talocalcaneal joint has three facets: Posterior (lateral), middle (medial), and anterior. The coalition usually occurs at the middle facet between the talus and sustentaculum tali. Ankylosis of the posterior or anterior facets is far less common.
   b. The talocalcaneal coalition is never seen on routine AP, lateral, and oblique films of the foot; initial diagnosis depends on secondary radiographic signs:
      (1) Talar beaking on lateral film: Dorsal

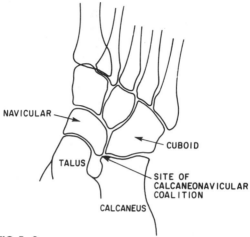

FIG 5-9.
Medial oblique view showing site of calcaneonavicular coalition.

      subluxation of the navicular is produced by subtalar rigidity, which leads to elevation of periosteum, which leads to subperiosteal proliferation and bony excrescence. Talar beaking is not pathognomonic; it is also seen in RA, diffuse idiopathic skeletal hyperostosis (DISH), and acromegaly.

    (2) Narrowing of the posterior subtalar joint on lateral film, representing degenerative joint disease (DJD) due to calcaneal eversion.

    (3) Ball and socket ankle seen on AP ankle (provides inversion-eversion function that is restricted at the talocalcaneal joint).

      c. Comparison views may be necessary to
evaluate for asymmetry of the
undersurface of the talar neck or
broadening of the lateral process of the
talus.

      d. Harris-Beath view (tangential calcaneus)
required but may give a false-positive or -
negative impression if patient is
positioned incorrectly (Fig 5–10).

D. Special techniques:

  1. Bone scan may be useful as a screening
device. In positive cases, the subtalar joint as
well as talar beak show increased uptake.

  2. Tomography: Lateral, complex motion; four
sections. Where W equals the heel width:

    a. $W-1$ cm demonstrates the medial facet.

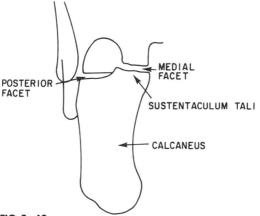

FIG 5–10.
Harris-Beath view of subtalar joint, demonstrating the
position and medial facets. The latter is the most likely
site of a talocalcaneal coalition.

b. W – 1.5 cm and W – 2 cm demonstrate the anterior facet.

c. W/2 demonstrates the posterior facet.

3. CT: Easily demonstrates the three facets of the subtalar joint with both feet viewed symmetrically in the same exam (Fig 5–11). Gantry either is not angled or is angled away from the knees. If mistakenly angled toward the knees, a false-positive impression may result.

4. Arthrography: Rarely performed today but

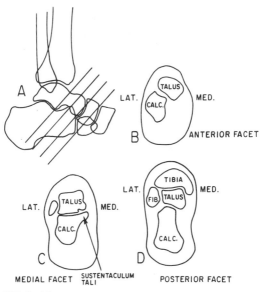

FIG 5–11.
CT for talocalcaneal coalition: **A,** position of cuts. **B,** anterior facet. **C,** medial facet. **D,** posterior facet.

can conclusively demonstrate a coalition whether it is osseous, cartilaginous, or fibrous.

XVI. CLEIDOCRANIAL DYSOSTOSIS

A. Definition: Autosomal-dominant abnormality with retarded development of the membranous bones.
B. Radiographic abnormalities:
   1. *Delayed cranial suture closure*; platybasia.
   2. *Wormian bones*.
   3. Partial or total *absence of the clavicle*, usually bilateral. Pseudoarthroses may be present.
   4. Ossification of the pubis absent or delayed. Valgus or varus deformities of the femoral neck.
   5. Various other epiphyseal or metaphyseal abnormalities may be present.

XVII. OSTEO-ONYCHODYSOSTOSIS (NAIL-PATELLA SYNDROME, OR FONG'S DISEASE)

A. Definition: Autosomal-dominant disorder characterized by multiple bone abnormalities, dysplastic fingernails, and renal disease.
B. Radiographic abnormalities:
   1. Absent patella.
   2. Posterior iliac horns are pathognomonic.
   3. Hypoplastic capitellum and radial head dislocation.
   4. Hypoplastic lateral femoral condyle and resultant genu valgum.

XVIII. CAUDAL REGRESSION SYNDROME (SACRAL AGENESIS)

A. Definition: Spectrum of abnormalities ranging from agenesis of part of the sacrum to agenesis of the sacrum, lumbar spine, and lower part of the thoracic spine. Twenty percent of patients are infants of diabetic mothers.

XIX. MADELUNG'S DEFORMITY

A. Definition: Bowing of the distal radius in an ulnar and volar direction. The ulna is therefore relatively long and is often dorsally dislocated. The deformity results in a narrowed carpal angle. Madelung's deformity may be seen:
   1. As an isolated finding.
   2. In dyschondro-osteosis dwarfism.
   3. In Turner's syndrome.
   4. As a deformity after radial fracture.
B. More common in women than men.

XX. CHROMOSOME DISORDERS

A. Down's syndrome (trisomy 21).
   1. Flared iliac wings with flattened acetabular roofs.
   2. Clinodactyly.
   3. Eleven ribs.
   4. Microcephaly.
   5. Atlantoaxial instability.
B. Trisomy 18.
   1. Adducted thumb and other finger and toe deformities.
   2. "Rocker-bottom" feet.
   3. Sternal aplasia or hypoplasia.
   4. Hypoplastic ribs and clavicles.
C. Turner's syndrome (deletion of one X chromosome).
   1. Short stature.
   2. Osteoporosis.
   3. Short metacarpals, especially the fourth.
   4. Madelung's deformity.
   5. Flattening of the medial tibial plateau with overgrowth of the medial femoral condyle.
   6. Cubitus valgus.

XXI. DWARFISM

by Paula Rand, M.D.

> **Key Concepts:** Differentiation of dwarfism is important for prognosis and genetic counseling. A standard set of films and an algorithm is presented that we have found useful in diagnosing different types of dwarfism.

Current research is revealing the modes of inheritance, life expectancy, and expected quality of life of the various dwarfs. Correct classification, based in large part on the radiographic findings, must be made to insure appropriate genetic counseling as well as appropriate transmission of information to parents regarding their child's life expectations and limitations.

Volumes have been written on the subject of dwarfism. The classifications and subclassifications change constantly as knowledge increases. The numerous synonyms and eponyms in use for clinical entities add to the confusion. The latest generally accepted classification was established by the European Society for Pediatric Radiology and the National Foundation March of Dimes; it is titled *International Nomenclature of Constitutional Diseases of Bones*. In this section, the eleven most common dwarfism complexes are described. This is followed by a simplified approach to classifying the different types of dwarfs.[5–11]

Selected Major Dwarf Syndromes

Achondroplasia

A. Synonyms: Chondrodysplasia, chondrodystrophia fetalis, chondrodystrophic dwarfism.

B. Inheritance: Autosomal-dominant (most cases,

however, are new mutants).
C. Age of manifestation: Birth.
D. Main clinical features:
   1. Short-limbed dwarfism with proximal segments of limbs shorter than distal (rhizomelic).
   2. Macrocrania, with or without hydrocephalus; characteristic facies with small maxillary area, prominent, bulging brow, and depressed nasal bridge; normal intelligence.
   3. Lumbar kyphosis in infancy progressing to lumbar lordosis in childhood and adulthood; occasional scoliosis; normal to relatively long trunk.
   4. Prominent buttocks and abdomen.
   5. Limitation of elbow extension; trident hand with relatively short fingers; genu varum.
   6. Possible neurologic symptoms due to relative spinal stenosis resulting from the narrowed interpedicular distance.
E. Major radiographic features:
   1. *Large calvaria with small area of facial bones* in comparison to vault; short base of skull and reduced size of the foramen magnum.
   2. *Decrease in the interpediculate distance from upper to lower lumbar spine*; short pedicles; *posterior scalloping of vertebral bodies*; occipitalization of C_1.
   3. *Squared iliac wings with horizontal acetabular roofs*; narrow greater sciatic notch; narrowed pelvic inlet.
   4. *Shortened and thickened tubular bones with flared metaphyses and normal epiphyses.*
   5. "Trident" hands in about 50% of cases.
F. Progression and Prognosis: Normal life span; complications occurring include hydrocephalus, nerve root irritation, and paraparesis and paraplegia.
G. Differential diagnosis: Other forms of short-limbed dwarfism:

1. Hypochondroplasia: Similar but milder changes; normal skull.
2. Pseudoachondroplasia: Normal skull; epiphyseal abnormality.
3. Thanatophoric dwarfism (infancy): Lethal condition; smaller thorax, more severe changes.

Hypochondroplasia

A. Synonyms: Chondrohypoplasia.
B. Inheritance: Autosomal-dominant; is distinct from achondroplasia and not a variation or manifestation of a single mutant gene.
C. Age of manifestation: Early childhood.
D. Main clinical features:
   1. Small stature with disproportionately short extremities.
   2. Slightly increased lumbar lordosis with mildly protuberant abdomen; kyphosis is not a feature.
   3. Limitation of elbow extension.
   4. Mild generalized joint laxity.
E. *Major radiographic features*:
   1. *Normal skull.*
   2. *Mild narrowing or lack of widening of the interpedicular distances from upper to lower lumbar spine.*
   3. *Tubular bones short and thickened with mild rhizomelic predominance.*
   4. *Large capital femoral epiphysis with short, broad femoral neck; prominent femoral trochanters.*
   5. Often *increased length of fibula.*
F. *Progression and prognosis*: Normal life span. Caesarean section for delivery is necessary because of the small pelvis.
G. *Differential diagnosis*:
   1. Achondroplasia: Skull involvement is present; has greater disproportion of limb length (rhizomelic); trident hand deformity present in about 50% of cases, more prominent

narrowing of interpediculate distance.
2. Metaphyseal chondrodysplasia, Schmid type: Pronounced coxa vara; growth plate line is vertical and irregular; no vertebral abnormalities.
3. Dyschondrosteosis: Madelung-type deformity present.

Pseudoachondroplasia

A. Synonyms: Pseudoachondroplastic type of spondyloepiphyseal dysplasia.
B. Inheritance: Probably heterogeneous; some are autosomal-dominant, others recessive.
C. Age of manifestation: Second to fourth year of life.
D. Main clinical features:
1. Normal face and head.
2. Short-limbed dwarfism but no rhizomelic predominance.
3. Marked joint laxity at all joints except elbow.
4. Short, stubby hands and feet.
E. Major radiographic features:
1. *Grossly abnormal epiphyses* with irregular calcification.
2. Normal skull.
3. *Variable deformities of the vertebrae* on the lateral view ranging from near normal to persistent oval shape to anterior beaking and platyspondyly; may have irregular ossification at the end-plates of the vertebral bodies.
F. Progression and prognosis: Normal life expectancy; early disability from osteoarthritis.
G. Differential diagnosis:
1. Achondroplasia: Head is large with frontal bossing and depression of bridge of nose; epiphyses are normal.
2. Multiple epiphyseal dysplasia: Vertebral anomalies are not present.

Thanatophoric Dwarfism

A. Synonyms: Severe achondroplasia.

B. Inheritance: Unknown; autosomal-recessive has been suggested.

C. Age of manifestation: At birth or in utero.

D. Main clinical features:
1. Death in early infancy.
2. Disproportionate dwarfism with marked rhizomelic tubular bone shortening.
3. Relatively normal trunk length, narrow thorax, protuberant abdomen.
4. Large head with frontal bossing, protruding eyes, and depressed nasal bridge.

E. Major radiographic features:
1. *Cloverleaf deformity of skull* may be present; short skull base.
2. *Platyspondyly* with notchlike ossification defects of the central portion of the upper and lower vertebral end-plates causing an H or U configuration. *Trunk length is normal due to widened disc spaces.*
3. *Lumbar interpedicular distance fails to widen caudally;* narrowing is greatest in midlumbar area.
4. *Short ribs* with consequent narrow thorax in AP and lateral views.
5. Decreased height and horizontal inferior margins of the iliac bones with small sacrosciatic notch.
6. *Marked rhizomelic tubular bone shortening with curving and irregular metaphyseal flaring; "telephone receiver-shaped" femurs.*

F. Progression and prognosis: Death within first few days of life due to respiratory distress and cardiac failure.

G. Differential diagnosis:
1. Achondroplasia: Fails to show the marked flattening of the vertebral bodies, widened intervertebral spaces, or the irregular flaring of the metaphyses.
2. Achondrogenesis: Lacks the telephone receiver-shaped femur; no narrowing of

sacrosciatic notch; lacks ossification of lower
vertebral bodies, iliac, and sacrum.
3. Metatropic dwarf: Decreased trunk length.

Achondrogenesis (Types I and II)
A. Synonyms: None.
B. Inheritance: Autosomal-recessive.
C. Age of manifestation: At birth or in utero.
D. Main clinical features:
1. Incompatible with life.
2. Extremely short limbs, protuberant abdomen,
disproportionately large head.
3. Hydropic, edematous appearance.
E. Major radiographic features:
1. Poorly mineralized skull.
2. *Ossification of spine absent or severely retarded
more caudally than cranially in spine.*
3. Absent ossification of talus and calcaneus.
4. *Barrel-shaped thorax with short, horizontal ribs*
with multiple fractures.
5. Very *small, broadened long bones* with concave
ends and longitudinally projecting spines or
spikes.
6. *Type II is less severe* with more complete
ossification and no evidence of rib fracture.
F. Progression and prognosis: Patients are stillborn
or die in neonatal period.
G. Differential diagnosis: All lethal forms of
dwarfism. Thanatophoric dwarfs demonstrate
markedly bowed femurs; ossification of vertebral
bodies, sacrum, and pubic and ischial bones is
present.

Asphyxiating Thoracic Dysplasia
A. Synonyms: Asphyxiating thoracic dystrophy,
Jeune syndrome, thoracic-pelvic-phalangeal
dystrophy, infantile thoracic dystrophy.
B. Inheritance: Autosomal-recessive.
C. Age of manifestation: Usually birth.
D. Main clinical features:
1. Respiratory difficulties due to small thoracic

cage.

2. Short extremities at birth, usually acromelic but sometimes rhizomelic.
3. Renal disease, progressive with age.

E. Major radiographic features:
   1. *Normal skull and spine.*
   2. *Small thoracic cage with short, horizontal ribs;* with time this becomes normal.
   3. *Sometimes flared ilia;* spurlike downward projections at the medial and lateral aspects of the acetabular roof; normalization of the pelvis with age.
   4. Short middle and distal phalanges with some coned epiphyses and premature fusion of epiphyses.
   5. *Disproportionately (usually acromelic) short extremities with metaphyseal irregularities.*

F. Progression and prognosis: If the patient survives the respiratory difficulties in the first year of life, there may be no further complications. Renal failure may develop in later childhood or adult life.

G. Differential diagnosis:
   1. Chondroectodermal dysplasia: Hair and teeth abnormalities.
   2. Thanotophoric dwarfism: Very short long tubular bones with characteristic bowing of the femurs.
   3. Achondroplasia: Thoracic cage is only mildly reduced in size, if at all; middle and distal phalanges are normal.

Chondroectodermal Dysplasia

A. Synonyms: Ellis-van Creveld syndrome; ectodermal dysplasia; chondrodystrophy with ectodermal defects; mesoectodermal dysplasia.
B. Inheritance: Autosomal-recessive.
C. Age of manifestation: Birth.
D. Main clinical features:
   1. Disproportionate dwarfism with short limbs,

particularly in the distal portions.
2. Polydactyly of the hand and often of the feet; extranumerary digit is on the ulnar aspect.
3. Sparse hair and disordered eruption or absence of teeth due to ectodermal dysplasia; hypoplastic nails.
4. Long, narrow thorax.
5. Congenital heart disease.

E. Major radiographic features:
1. *Normal skull and spine.*
2. *Short ribs and long narrow thorax in infancy,* progressing to normal as child grows.
3. *Long bones are markedly short,* particularly the distal segments.
4. *Polydactyly* with or without fusion of metacarpals and/or phalanges.
5. In infancy, dysplasia of the pelvis with small iliac bones and, sometimes, a downward-directed spike in the region of the triradiate cartilage. This, too, progresses to normal with age.
6. *Fusion of capitate and hamate;* coned epiphyses of middle and distal phalanges.
7. *Proximal tibial epiphyses hypoplastic and displaced medially.* Defect of the lateral tibia causes severe genu valgum.

F. Progression and prognosis: Prognosis is related to the congenital heart defect with infant mortality estimated to exceed 50%. Also, the long, narrow thorax produces pulmonary insufficiency. Some disability may result from the genu valgum.

G. Differential diagnosis:
1. Asphyxiating thoracic dystrophy in neonatal period; hypoplastic nails are associated, and polydactyly is inconstant.
2. Later, other forms of short-limbed dwarfism, but upper tibial deformity and fused capitate and hamate should distinguish chondroectodermal dysplasia.

Diastrophic Dwarfism (Dysplasia)

A. Synonyms: None.
B. Inheritance: Autosomal-recessive.
C. Age of manifestation: Birth.
D. Main clinical features:
1. Small stature, short extremities; clubfoot and hitchhiker's thumb.
2. *Contractures of many joints.*
3. Cystic swelling of external ear.
E. Major radiographic features:
1. *Normal skull.*
2. *Appearance of spine variable*: Spectrum ranges from flattening to some increase in height, but may be normal. Posterior scalloping may be present.
3. Short clubbed long bones; epiphyses may be flattened and stippled.
4. *Severe talipes equinovarus* with short, thick metatarsals.
5. *Short first metacarpal with abnormal origin of thumb.*
6. Scoliosis.
F. Progression and prognosis: Perinatal mortality is increased, but those who survive past infancy can expect a normal life span. Moderate to severe restriction of physical activity due to combination of severe clubfoot deformity and scoliosis. Osteoarthritis is inevitable.
G. Differential diagnosis:
1. Arthrogryposis multiplex: Marked decrease in muscle mass; lack of characteristic skeletal changes in hands and feet.
2. Achondroplasia: Skull will be affected; contractures are not a feature and epiphyses are normal.

Chondrodysplasia Calcificans Punctata

A. Synonyms:Chondrodysplasia punctata, Conradi's disease, congenital stippled epiphysis, Conradi-Hunerman disease, chondropathia calcificans, chondropathia punctata, chondroangiopathia punctata, dysplasia epiphysealis congenita, punctate epiphyseal dysplasia.

B. Inheritance: Two types, both autosomal-dominant and autosomal-recessive.

C. Age at manifestation: Birth or early infancy.

D. Main clinical features:
1. Congenital short-limbs, short stature; may be asymmetric (dominant type) or disproportionate short stature of the rhizomelic type (recessive type).
2. Flat face with depressed bridge of nose.
3. Ichthyosiform skin changes; joint contractures; cataracts.
4. Associated anomalies include congenital heart disease (dominant type).

E. Major radiographic features:
1. Dominant type (Conradi):
   a. Infancy:
      (1) *Stippling of epiphyses,* primarily at the ends of long bones, carpal, and tarsal regions.
      (2) *Unilateral, less frequently bilateral, shortening of the long tubular bones.*
      (3) Coronal cleft vertebral bodies; scoliosis may be present.
   b. Childhood and adulthood:
      (1) *Assymmetric, less frequently symmetric, shortening of the long tubular bones* and metacarpal bones.
      (2) *Epiphyseal dysplasia or irregularity* in places where previous stippling was located.
2. Recessive type:
   a. *Severe symmetric shortening and metaphyseal splaying and calcific stippling of the ends of humeri and/or femurs.*
   b. *Dorsal and ventral ossification centers of vertebral bodies, which are separated by radiolucent bars of cartilage.*

F. *Progression and prognosis:*
1. Dominant type: If severely affected, patients may be stillborn or die within the first week of life. After the first week, life expectancy and intellectual development are normal. Complications arise from cataracts, joint contractures, and other orthopedic problems.
2. Recessive type: Prognosis is extremely poor. Failure to thrive, severe retardation in psychomotor development and death in infancy from repeated infection are demonstrated.

G. *Differential diagnosis:*
1. Dominant type:
   a. Recessive type: Coronal clefts in vertebral bodies and symmetric shortening of the limbs.
   b. Zellurger's syndrome: Associated clinical, cytologic, and biochemical findings differentiate it.
   c. Epiphyseal dysplasia.
2. Recessive type:
   a. Dominant type: Vertebral bodies lack coronal clefts; more often asymmetric.
   b. Epiphyseal dysplasia.

Dyschondrosteosis

A. Synonyms: Leri-Weill disease, Madelung's deformity with short forearms.
B. Inheritance: Autosomal-dominant with female preponderance.
C. Age of manifestation: Later childhood.
D. Main clinical features:
1. Short-limbed, short stature associated with forearm and tibia and fibular shortening.
2. Dorsal dislocation of distal ulna with short, bowed radius.
3. Limitation of motion at the elbow and wrist.
E. Major radiographic features:
1. *Normal skull, spine, thorax,* shoulder girdle/

humerus, *pelvic girdle,* and *femur.*
2. *Dorsal dislocation of ulna at wrist* and sometimes at elbow.
3. *Shortening of radius with bowing* and increase in distance between distal radioulnar articulation.
4. *Pyramidal appearance of the carpus with the lunate at the apex,* fitting between the radius and ulna.

F. Progression and prognosis: Normal life expectancy. Some limitation of wrist motion, occasionally accompanied by pain.

G. Differential diagnosis:
1. Other mesomelic dwarfs: Madelung's deformity is not present.
2. Turner's syndrome: Clinical manifestations should easily differentiate.

Spondyloepiphyseal Dysplasia Congenita

A. Synonyms: None.
B. Inheritance: Autosomal-dominant with considerable variability of expression; may be sporadic.
C. Age of manifestation: Birth.
D. Main clinical features:
1. Markedly reduced stature with short trunk.
2. Pectus carinatum.
3. Gross coxa vara.
4. Decreased muscle tone and waddling gait.
5. Myopia and retinal detachment.
6. Occasionally clubfoot and/or cleft palate.

E. Major radiographic features:
1. *Flattening of vertebral bodies* with shortening of the trunk; *hypoplasia of the odontoid.*
2. *Small ilium* with irregular acetabula with horizontal roofs; *delayed ossification of pubic rami.*
3. Capital femoral epiphyses, if present, are small and deformed with delayed presentation.

4. *Limb epiphyses vary from normal to severely fragmented. Metaphyses also may be disordered and irregular.*
5. *Shortening of the long bones.*
6. Scoliosis.

F. Progression and prognosis:
   1. Premature osteoarthritis in affected joints is inevitable.
   2. Scoliosis is probable, and paraplegia has occurred.

G. Differential diagnosis:
   1. Morquio's disease: Clinically similar but radiographically dissimilar.
   2. Other forms of spondyloepiphyseal dysplasias: Most have less severe coxa vara, lack the delayed ossification of the pelvis and proximal femur, and do not demonstrate major changes in hands and feet.

Approach to Radiographic Classification of Dwarfism

Radiographic evaluation should be performed as early in life as possible since many of the typical findings become less apparent as the patient matures. A specific set of radiographs should be obtained in all evaluations and should include the following:

A. Lateral skull.
B. AP and lateral of the thoracolumbosacral spine.
C. Chest (shoulders must be included).
D. AP pelvis and hips.
E. AP upper extremity (one side only).
F. AP lower extremity (one side only).
G. PA hand/wrist (one side only).

Initially one should determine what feature constitutes the most striking abnormality. Fig 5–12 is a flow chart that should allow a physician to accurately characterize most dwarfs. Tables 5–1 through 5–6 give the most likely diagnoses once the flow chart has been consulted. Each of these tables lists only the three most

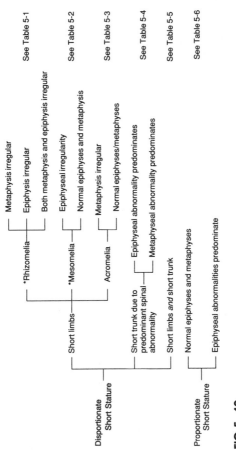

FIG 5–12.
Flow chart for characterization of most types of dwarfism. Asterisk indicates conditions for which tables of limb ratios are available for subtle or questionably abnormal cases.[6]

TABLE 5–1.

Group 1: Disproportionate Short Stature

A. Short-limbed, normal trunk
  1. *Rhizomelia* (proximal limbs more affected)
    a. Metaphysis irregular or abnormal
      (1) Achondroplasia
        a Macrocrania with short skull base and small foramen magnum
        b Decreasing interpedicular width from upper to lower lumbar spine
        c Squared iliac wings with narrow sacrosciatic notch
      (2) Hypochondroplasia
        a Normal skull
        b Mildly decreasing interpediculate distances from upper to lower lumbar spine
        c Often increased length of fibulas
      (3) Achondrogenesis
        a Lack of ossification of lower lumbar and sacral spine
        b Disproportionately large head
        c Marked micromelia
      (4) Thanatophoric dwarfism
        a Marked shortening, curving of tubular bones with metaphyseal flaring
        b Platyspondyly; H- or U-shaped vertebrae
        c Narrow thorax both AP and lateral views due to rib shortening
    b. Epiphysis irregular or abnormal
      (1) Chondrodysplasia calcificans punctata (recessive)
        a Symmetric rhizomelic shortening of extremities
        b Coronal clefts of vertebral bodies
        c Calcific stippling ends of humeri and/or femurs

(Continued.)

TABLE 5–1 (*cont.*).

   c. Both epiphyses and metaphyses are
      abnormal
      (1) Pseudoachondroplasia
         a Normal face and head
         b Grossly abnormal epiphyses
         c Variable deformities of vertebrae

common radiographic abnormalities. Because of this
limitation and the fact that most dwarfism syndromes
have a spectrum of involvement that may or may not
include the most common radiographic findings, one
may also have to refer back to the more complete
descriptions or use the excellent references listed at the
end of this chapter for further clarification. In difficult

TABLE 5–2.

Group 1: Disproportionate Short Stature

A. Short-limbed, normal trunk
   2. *Mesomelia*
      a. Epiphyses irregular or abnormal
         (1) Chondroectodermal dysplasia
            a Short-limbed dwarfism most
               pronounced distally
            b Postaxial polydactyly
            c Fusion of capitate and hamate
      b. Normal epiphyses and metaphyses
         (1) Dyschondrosteosis
            a Normal skull, spine, thorax
            b Dorsal dislocation of ulna at wrist with
               Madelung's deformity
            c Pyramidal appearance of the carpus
               with the lunate at the apex

TABLE 5–3.

Group 1: Disproportionate Short Stature

A. Short-limbed, normal trunk
   3. *Acromelia*
      a. Normal metaphyses and epiphyses
         (1) Acrodysostosis
            a Short stature predominantly distal, with peripheral dysostoses
            b Occasional brachycephaly and thickening of vault and base
            c Nasal hypoplasia
      b. Irregular or abnormal metaphyses
         (1) Asphyxiating thoracic dysplasia
            a Small thorax with short, horizontal ribs
            b Short limbs with acromelic predominance and metaphyseal irregularity
            c Short middle and distal phalanges; coned epiphyses

---

cases starting over on the flow chart with the second most striking abnormality may prove more helpful in reaching the correct diagnosis.

You will note that several types of dwarfs appear in more than one table. The explanation is threefold. First, for many types, there is a wide spectrum of findings with certain of the findings present in one individual and not in another. Second, the age at evaluation will affect the abnormality most predominant at presentation because with growth, certain characteristic abnormalities either resolve or become more dominant. Third, as more is learned about dwarfs with individual characteristic manifestations, more distinct entities are described. In the earlier literature, some of these same entities would have been considered together, so there is some disagreement among the resources regarding some of the characteristic findings.

Finally, good luck! Although the task of classifying a dwarf is tedious and often quite frustrating, the final diagnosis is very important to the patient and his or her family—for genetic counseling and a better understanding of life's expectations and limitations.

XXII. MUCOPOLYSACCHARIDOSES

A. Definition: *Spectrum* of hereditary diseases characterized by a distinctive dwarfism pattern and distinguished in part by clinical presentation and largely by the difference in

TABLE 5–4.

Group 1: Disproportionate Short Stature

B. Short trunk and major spinal involvement
  1. Epiphyseal abnormalities predominate
    a. Spondyloepiphyseal dysplasia
      (1) Platyspondyly with hypoplastic odontoid
      (2) Shortening of long bones with epiphyseal, and less often metaphyseal, abnormality
      (3) Coxa vara
    b. Chondrodysplasia calcificans punctata (dominant type).
      (1) Asymmetric shortening of limbs
      (2) Scoliosis with irregular deformities of vertebral bodies
      (3) Calcific deposits in and around epiphyses and other cartilaginous areas
  2. Metaphyseal abnormalities predominate
    a. Spondylometaphyseal dysplasia
      (1) Platyspondyly with mild irregularity of end-plates
      (2) Metaphyseal irregularities with almost normal epiphyses

TABLE 5–5.

Group 1: Disproportionate Short Stature

C. Short limbs and short trunk: All listed have both metaphyseal *and* epiphyseal abnormalities, so this focus will not help to differentiate
  1. Diastrophic dysplasia (dwarf)
     a. Severe talipes equinovarus with short, thick metatarsals
     b. Short first metacarpals with abnormal origins of thumbs
     c. Scoliosis
  2. Pseudoachondroplasia
     a. Normal face and head
     b. Variable deformities of the vertebrae
     c. Grossly abnormal epiphyses
  3. Kniest's disease
     a. Large cranium; hypoplasia of odontoid; flat face with depressed nasal bridge
     b. Severe platyspondyly
     c. Flared metaphyses; large, irregular, punctate epiphyses
  4. Metatropic dwarfism
     a. Kyphoscoliosis
     b. Striking platyspondyly
     c. Diaphyseal constriction and widely flaring metaphyses

---

mucopolysaccharides excreted in the urine; all but Morquio's patients are *mentally retarded*.
B. *Radiographic findings common to all the mucopolysaccharidoses:*
  1. *Macrocephaly*, with a *J-shaped sella*.
  2. *Ribs are oar shaped* (i.e., narrow at costovertebral junction, then broad).
  3. Lateral view of the thoracolumbar spine is distinctive. There is *a focal kyphosis at the*

TABLE 5–6.

Group 2: Proportionate Short Stature

A. Normal epiphyses and metaphyses
  1. Systemic diseases/metabolic abnormalities (most are addressed in other chapters):
     a. Pituitary dwarfism
     b. Renal osteodystrophy
     c. Gonadal dysgenesis
     d. Cardiopulmonary disease
     e. Regional enteritis
     f. "Hepatic" dwarf (biliary atresia or cirrhosis)
     g. Hematologic abnormality (severe anemias)
     h. Mucosaccharidosis
     i. Mucolipidosis
     j. Other storage diseases
     k. Malnutrition
B. Predominant epiphyseal abnormalities
  1. Multiple epiphyseal dysplasia
     a. Irregularity of end-plates in lower dorsal spine
     b. Pair of joints almost always symmetrically involved; fragmented ossification centers
     c. Capital femoral epiphysis almost always involved
  2. Chondrodysplasia punctata dominant (Conradi-Hünermann)
     a. Stippled epiphyses
     b. Mild shortening to normal limb length; if abnormal, usually asymmetric
     c. Irregular deformities of the vertebral bodies
  3. Hypothyroidism
     a. Broad provisional zones of calcification
     b. Retarded skeletal and skull maturation
     c. Stippled epiphyses, especially femoral capital epiphyses, which has been termed "cretinoid" and resembles Legg-Calve-Perthes disease

(Continued.)

TABLE 5–6 (cont.).

C. Predominant metaphyseal abnormalities
  1. Metaphyseal chondrodysplasias
    a. Vertebral bodies may retain their oval shape
    b. Limb metaphyses expanded and irregular; may be coned
    c. Femoral metaphyses (proximal) demonstrate medial beaking, irregularity and widening; develops coxa vara
  2. Spondylometaphyseal dysplasia
    a. Universal platyspondyly; mild irregularity of the end-plates
    b. Almost normal epiphyses; metaphyseal irregularities
    c. Marked coxa vara may develop

---

    thoracolumbar junction, with an L_1 *or* L_2 *body that is distinctly small, retrolisthesed, and oval in shape with a central anterior beak.*
  4. *AP pelvis demonstrates constricted iliac necks and flared iliac wings.*
  5. *Long bones are shortened,* often with wide metaphyses and diaphyses.
  6. PA view of hands demonstrates *short wide metacarpals,* with constricted proximal ends. The overall appearance is that of a *fan-shaped configuration of the metacarpals.*
C. *Morquio's* syndrome: Treated separately since these patients are *not mentally retarded.* In addition to the osseous abnormalities seen in other mucopolysaccharidoses they have the following distinctive osseous findings:
  1. *Severe platyspondyly* with anterior beaking of the vertebral bodies.
  2. *Hypoplastic odontoid.*
  3. *Atlantoaxial instability.*

4. Compression and fragmentation of the femoral capital epiphyses.
5. Restricted chest wall motion secondary to sternal buckling.

REFERENCES

1. Gold R, Amstutz H: Surgical procedures for congenital dislocation of the hip. *Radiol Clin North Am* 1975; 13:123–137.
2. Morin C, Horcke H, MacEwan G: The infant hip: Real-time US assessment of acetabular development. *Radiology* 1985; 157:673–677.
3. Ozonoff M: *Pediatric Orthopedic Radiology.* Philadelphia, W. B. Saunders, 1979.
4. Freiberger R, Hersh A, Harrison M: Roentgen examination of the deformed foot. *Semin Roentgenol* 1970; 5:341–353.
5. Bailey JA II: *Disproportionate Short Stature—Diagnosis and Management.* Philadelphia, W. B. Saunders, 1973.
6. Felson B (ed): Dwarfs and other little people: A roentgen guide. *Semin Roentgenol* 1973; 8:
7. Greenfield GB: *Radiology of Bone Diseases, ed 4.* Philadelphia, J. B. Lippincott, 1986.
8. Spranger JW, Langer LO Jr, Wiedemann HR: *Bone Dysplasias: An Atlas of Constitutional Disorders of Skeletal Development.* Philadelphia, W. B. Saunders, 1974.
9. Robinow M, Chumlea WC: Standards for limb bone length ratios in children. *Radiology* 1982; 143:433–436.
10. Tayki H: *Radiology of Syndromes and Metabolic Disorders, ed 2.* Chicago, Year Book Medical Publishers, 1983.
11. Wynne-Davis R, Hall CM, Apley AG: *Atlas of Skeletal Dysplasias.* New York, Churchill Livingstone, 1985.

6 | Miscellaneous, Including Hematologic Disorders and Infection

I. HEMATOLOGIC DISORDERS
A. Hemophilia

> *Key Concepts:* Males only are affected; hemarthroses may appear dense; deformities and contractures; arthropathy in knees, elbows, ankles; pseudotumor in femur, pelvis.

A. Definition: A bleeding disorder due to a clotting factor deficiency that results in hemarthroses, deformities, and arthropathy.
B. Epidemiology: The 2 most common varieties are inherited through an X-linked recessive pattern and are, therefore, found only in *males*:
   1. *Hemophilia A (factor VIII).*
   2. *Hemophilia B (factor IX; Christmas disease).*
   3. *von Willebrand's disease* is due to a combined factor VIII deficiency and platelet abnormality and is found in *males or females*.
C. Radiographic abnormalities:
   1. *Hemarthroses:*
      a. Often lead to *flexion contractures* and

arthropathy.

b. May involve multiple joints, but often *asymmetric.*

c. Joints most commonly involved are *knee, elbow, ankle,* hip, shoulder (in descending order of occurrence).

2. Arthropathy: Follows multiple episodes of hemarthroses.

a. *Effusions appear dense* radiographically due to *hemosiderin deposits* in hypertrophied synovium.

b. The synovial inflammation causes *hyperemia,* which in turn causes *osteoporosis, epiphyseal overgrowth* (seen as *flared, enlarged joints,* with gracile diaphyses), and *early epiphyseal fusion* (and resultant short bones).

c. *Cartilage degeneration, erosions,* and *subarticular cysts* are seen uniformly throughout the joint.

d. *Secondary degenerative joint disease (DJD)* eventually develops.

e. Distinctive findings in the *knee* include *widening of the intercondylar notch* and *squaring of the inferior pole of the patella,* but they are not pathognomonic since juvenile rheumatoid arthritis (JRA) may have the same appearance.

f. Distinctive findings in the *elbow* include an *enlarged radial head* and *enlarged trochlear notch.*

g. Distinctive findings in the ankle include a nonspecific tibiotalar slant.

3. Pseudotumor.

a. Due to an *intraosseous, subperiosteal, or soft tissue bleed.*

b. Occurs most commonly in the *femur, pelvis,* and *tibia.*

c. Most commonly appears as *extrinsic and/or intrinsic scalloping and pressure erosion.* The

area of destruction may be extremely large, *but* the *margins are generally sclerotic and sharp.*
   d. There *may be a large soft tissue mass,* depending on the site of origin of the bleed.
   e. *Periosteal reaction* may be extensive.
   f. The size and extent of destruction may simulate neoplasm, but sclerotic margins with both extrinsic and intrinsic scalloping suggest the correct diagnosis.

B. Anemias

> *Key Concepts:* Sickle cell—dactylitis, avascular necrosis (AVN), infarcts, infections; thalassemia— more severe marrow hyperplasia with squared phalanges and "hair-on-end" appearance on skull.

A. Definition: The congenital anemias represent abnormalities in one of the chains comprising hemoglobin which affect the shape and/or function of hemoglobin.
B. Radiographic abnormalities: The following skeletal abnormalities may be found in each of the anemias, but each may be more predictably seen with one in particular.
   1. *Marrow hyperplasia* due to long-term anemia: *Osteopenia,* but *coarsened trabeculae, widening of tubular bones and mandible, widened diploic spaces* in the skull.
   2. *Infarction* due to vascular occlusion: *Dactylitis* (hand-foot syndrome) due to infarcts in the small tubular bones of the hands and feet, where hematopoietic marrow is persistent and digits are unprotected from *vasoconstriction due to ambient cold. Periosteal reaction and soft tissue swelling occur,* and the entity is *often confused with osteomyelitis.*
      a. *Diaphyseal infarcts*: Patchy sclerosis or

serpiginous calcified density with
occasional periosteal reaction.
   b. *Vertebral body end-plate infarcts*: Pattern is
different, depending on type of anemia
(see C, below).
   c. *AVN*: Especially femoral and humeral
heads; often bilateral.
3. Predilection for *osteomyelitis*, often
indistinguishable from infarction.
Staphylococcus is the most common organism,
but *Salmonella* osteomyelitis is seen much
more often in the anemias than in the normal
population.
C. Features of the various anemias.
   1. *Sickle cell (Hb SS)*:
      a. Found in 1% of North American blacks.
      b. *Dactylitis* is very common (10% to 20% of
children with sickle cell disease).
      c. *AVN* is extremely common.
      d. Necrosis of *vertebral body* end-plates is
described as *H-shaped* and is due to the
distribution of small vessel arcades in
which the sickled cells "sludge."
      e. Other radiographic findings include *renal
papillary necrosis, cholelithiasis, splenic
infarction, cardiomegaly,* and *pulmonary
infarction.*
   2. Sickle cell trait (Hb AS): Very few
musculoskeletal findings, only an occasional
bone infarct.
   3. Sickle cell hemoglobin C (Hb SC): Marrow
hyperplasia of the skull and avascular
necrosis predominate; splenomegaly.
   4. *Thalassemia* (Cooley's anemia).
      a. Thalassemia major manifests early, and
death usually occurs by young adulthood.
      b. *Marrow hyperplasia* is spectacular, with
dense striations in a very widened diploic
space (hair-on-end); obliteration of
paranasal sinuses; Erlenmeyer flask

deformity; squared phalanges.
c. AVN much less common than in sickle cell disease.
5. Sickle cell thalassemia: Ranges from asymptomatic to typical sickle cell manifestations (infarcts overshadow hyperplastic marrow changes).

II. INFECTION

> *Key Concepts:* Osteomyelitis may appear extremely aggressive, simulating neoplasm. Soft tissue planes often are obliterated. Osteomyelitis in infants may involve the epiphyses. It is metaphyseal in children; radiographic changes lag behind clinical findings. Septic joints must be proven emergently by aspiration.

A. *Organisms.*
1. Overall, *Staphylococcus aureus* is by far the most common organism.
2. In *neonates, group B Streptococcus* is more common.
3. *Drug abusers* have increased incidences of *Pseudomonas* and *Serratia* osteomyelitis compared to the rest of the population; even in this group, *Staphylococcus* is still the most common organism.
4. *Sickle cell* patients have an increased incidence of *Salmonella* osteomyelitis, but *Staphylococcus* is most common in this group as well.
5. Tuberculosis and syphilis, which are of radiographic interest as separate entities, are discussed in F1 and F2, below.
B. Osteomyelitis: *Routes of involvement.*
1. *Direct penetrating wound.*
2. *Contiguous spread* from soft tissues to periosteum and subsequent bone involvement. Some areas are particularly at

risk:

   a. Metacarpals and phalanges (as well as metacarpophalangeal [MCP] and proximal interphalangeal [PIP] joints) from a human bite (most often from punching an adversary in the mouth).

   b. A felon (infection in the terminal pulp of the digit) may progress to osteomyelitis of the tuft.

   c. In the hand, soft tissue infection may spread along tendon sheaths and fascial planes, so the site of bone involvement may be distant from an initial injury.

   d. Decubiti in bedridden or paraplegic patients often result in osteomyelitis, especially of the sacrum and ischial tuberosities.

   e. Ulcers on the feet of diabetics are discussed in F3, below.

   f. A stubbed toe with a nail bed injury may result in osteomyelitis of the distal phalanx of the hallux since the periosteum is immediately adjacent to the nail bed.

3. *Hematogenous spread*: As is the case with bone tumors, site of bone involvement (both site within the bone and the individual bone) is a major determinant in making the diagnosis of hematogenous osteomyelitis. The site of involvement is strongly influenced by vascular anatomy.

   a. *Infant* (neonate to 12 months): Some metaphyseal vessels penetrate the epiphyseal plate to anastomose with epiphyseal vessels. *Metaphyseal infections, therefore, not uncommonly involve the epiphysis and joint* in this age group, *resulting in slipped epiphyses and growth deformity.* It is not uncommon to have *multifocal* sites of involvement in infants, often clinically benign, radiographically simulating metastatic neuroblastoma or an aggressive histiocytosis.

b. *Child: Terminal vessels* occur as loops with sluggish blood flow *in the metaphyses* and do not cross into the epiphyses. This, combined with a relative *lack of phagocytes* in the metaphyseal region, *leads to the metaphyses being the most common site of infection in the child. Epiphyseal and joint involvement are rare. The tubular bones of the lower extremities* are the most common sites (65% to 75% of childhood osteomyelitis occurs in the metaphyses of the femur or tibia).

c. *Adult*: With closure of the epiphyseal plate, the terminal metaphyseal vessels anastomose with epiphyseal vessels. Thus *joint involvement secondary to osteomyelitis is more common*. The *spine and small bones* are more commonly involved than tubular bones.

C. Osteomyelitis: Radiographic appearance varies depending on the clinical course of the infection.
   1. *Acute osteomyelitis*:
      a. Radiographic change lags behind onset of infection by 1 to 2 weeks. *First sign* is *blurring* or *obliteration of soft tissue fat planes*. This may be an important differential point since fat planes are often retained but displaced by soft tissue tumor.
      b. Soft tissue changes are followed by *intramedullary destruction*, often in an extremely subtle permeative pattern that may appear only as a focal decrease in density. If this permeative pattern is serpiginous, it assures the diagnosis of osteomyelitis.
      c. This is followed by *cortical destruction, endosteal scalloping*, and *periosteal reaction* (expected by 2 weeks).

d. These initial bone changes appear highly aggressive and may be difficult to differentiate from tumor. Knowing the time course of the disease may be helpful since an acute osteomyelitis may cause destructive changes more rapidly than does tumor.

e. Eventually, a *sequestrum and involucrum* may develop. A sequestrum is necrotic bone isolated from living bone by granulation tissue; it appears relatively dense because it has no blood supply while the surrounding bone is hyperemic and loses its mineralization. A sequestrum may harbor bacteria, leading to chronic osteomyelitis. An involucrum develops secondary to lifting of the periosteum by the abscess and subsequent reactive new bone formation and is living normal bone that envelopes the sequestrum. Involucrum formation is more common in children than adults since the periosteum is relatively loosely attached at the metaphysis, allowing easier periosteal elevation to initiate the formation of the host bone reaction.

2. Subacute osteomyelitis: *Brodie's abscess.*

a. Usually found in the *metaphysis of a child.*

b. A lucent focus of osteomyelitis *sharply delineated* by a *sclerotic margin.*

c. Therefore, appears much less aggressive than acute osteomyelitis.

d. Clinically it is also less distinctive, often without fever or elevated sedimentation rate.

e. The diagnosis may therefore be difficult, with the *differential* including *eosinophilic granuloma* and other benign metaphyseal lesions.

f. *Occasionally,* a Brodie's abscess is *cortically*

based, *eliciting more sclerosis* and periosteal
reaction. The *differential* diagnosis in this
situation may include *osteoid osteoma and
subacute stress fracture.*

3. *Chronic osteomyelitis.*
   a. *More host reaction,* often with thickened
      cortices and variable sites of lucency and
      density; *may have sequestra* present.
   b. *Plain film appearance may not change* for a
      period of several years, *yet may reactivate.*
   c. Evaluate by watching for change in
      osteolysis or development of periosteal
      reaction. If no radiographic change is seen
      in a patient who clinically suggests
      reactivation of chronic osteomyelitis,
      nuclear medicine studies are indicated (see
      D4, below).

D. Osteomyelitis: Diagnostic difficulties.
   1. Differentiation of osteomyelitis from
      aggressive lytic tumor may be difficult since
      the nature of destructive pattern or periosteal
      reaction is not pathognomonic. Soft tissue
      characteristics may be useful.
   2. Patients on partial antibiotic therapy may
      have delayed radiographic changes.
   3. Diagnosing infection in total knee
      arthroplasties (TKAs) is notoriously difficult
      since radiographic change is seldom seen.
      The physician should be very cautious about
      diagnosing absence of infection in TKA.
   4. Other modalities may be useful in the
      diagnosis of osteomyelitis and should be
      tailored to the individual needs of the case.
      a. Tomography may be useful in identifying
         sequestra for surgical resection.
      b. CT findings usually mirror those of the
         plain film: Destructive bone changes,
         sometimes with serpiginous tracking, are
         seen. Soft tissue fat planes are obliterated.
         Soft tissue abnormality often involves

several muscle groups and is less discrete than many soft tissue tumor masses. Contrast enhancement is usually nonuniform and often forms a swirling pattern through the soft tissue mass.

   c. Radionuclide studies.

     (1) ^{99m}Tc-MDP bone scans: In adults, highly sensitive in detecting early occult osteomyelitis; in children, perfectly symmetric position for side-to-side comparison, as well as pinhole collimators allow differentiation of metaphyseal osteomyelitis from the normal increased uptake in the metaphyseal-epiphyseal growth centers. Specificity is not as good as sensitivity in bone scans, but a 3- or 4-phase technique (perfusion, blood pool, bone uptake, and, sometimes, 24-hour images) increase the specificity. Increasing lesion-to-background ratios make osteomyelitis more likely than cellulitis.

     (2) Both ^{67}Ga and ^{111}In WBC imaging improve specificity (75% to 85%)[1, 2] but at the cost of delaying the diagnosis.

E. Complications of osteomyelitis.

  1. Chronic osteomyelitis with exacerbations.

  2. Slipped epiphyses and deformity.

  3. Joint involvement and eventual osteoarthritis.

  4. Amyloid deposit in chronic active lesion.

  5. Squamous cell carcinoma developing in a chronic draining sinus tract, usually 10 to 20 years following development of chronic osteomyelitis.

F. Special cases of osteomyelitis.

  1. *Tuberculosis (TB)*:

    a. Tends to have a *slower course* and *less host reaction* than pyogenic osteomyelitis. In

this appearance, it is similar to many fungal infections.
   b. TB dactylitis *(spina ventosa)* is an unusual manifestation seen in the tubular bones of the hands and feet, usually in children; may be multifocal. Radiographic characteristics are soft tissue swelling with periostitis followed by expansion of the bone. Differential diagnosis includes JRA, sickle cell dactylitis, and other infections.
2. *Syphilis* osteomyelitis.
   a. *Congenital*: Initially demonstrates *metaphyseal irregularity* and a widened zone of provisional calcification, occasionally resulting in slipped epiphyses. May progress to invade the diaphysis and elicit *periosteal reaction*. Congenital syphilis is in the extensive differential for infants with periosteal reaction, including nonaccidental trauma, tumor, other infections, and metabolic diseases.
   b. *Acquired* (not distinguishable from reactivated congenital syphilis): This chronic osteomyelitis elicits a periostitis and endosteal reaction which result in an *enlarged, bowed bone with mixed lytic and sclerotic areas*. In the tibia, the bow is anterior and is called a *saber shin* deformity. This differs from Paget's disease of the tibia since syphilis is diaphyseal and rarely involves the subarticular region of the bone. Flat bones and cranium may also be involved in syphilis osteomyelitis.
   c. Another manifestation of syphilis is neuropathic joints, especially knees (see Chapter 3).
3. Osteomyelitis in the diabetic foot.
   a. Ulcerations and underlying osteopenia

make the diagnosis of acute osteomyelitis difficult since early findings of osteomyelitis consist of soft tissue abnormality and decreased bone density. Therefore, there is heavy reliance on periosteal reaction, cortical destruction, and progression on serial films.

   b. Diabetics also often have neuropathic joints (most commonly the talonavicular and tarsometatarsal joints). This process results in cortical destruction, fragmentation, loss of cartilage, and effusion—findings that, in other joints, are suggestive of infection.

   c. Radionuclide studies may be very useful (see D4, above).

4. *Pin tract osteomyelitis.*

   a. Suspect with an enlarging lucency adjacent to a pin and associated periosteal reaction.

   b. If a lucency about the pin is due only to motion of the pin, the lucency is usually uniform and well-marginated.

   c. A *ring sequestrum* (a dense ring of sclerotic bone surrounded by irregular lucent destruction) is diagnostic of osteomyelitis in a pin tract.

G. *Septic arthritis.*

1. Radiographic signs in bacterial septic arthritis:

   a. *Joint effusion* is first sign.

   b. Hyperemia leads to *osteoporosis.*

   c. Cartilage destruction with *decreasing joint space.*

   d. *Bone erosion* and destruction follow rapidly.

   e. Osteomyelitis may develop via contiguous spread.

   f. *Sclerotic* host *reaction.*

   g. Ankylosis may occur eventually.

2. *TB* and *fungal septic arthritis.*
   a. *Little* or no host bone *reaction.*
   b. *Cartilage destruction* is much *slower.*
   c. *Slow progression of erosions,* often without other abnormalities.
   d. Hip and knee are most common areas of involvement, followed by wrist and elbow.
3. Radionuclide studies may be helpful but joint aspiration more quickly assures the diagnosis. Note that a prior aspiration, with or without the use of radiographic contrast material, does not alter bone scan results.[3]

H. Special considerations in diagnosing septic arthritis.
   1. *Hip in childhood.*
      a. Common site of septic arthritis since a hematogenous focus of infection in the metaphysis is within the hip joint capsule.
      b. Soft tissue signs of effusion (bulging psoas, gluteal, and obturator internus fat planes) are of limited usefulness since they may be false-positive with any pelvic rotation or flexion or rotation of the hip.
      c. Inducing an *air arthrogram* with traction on the hip rules out an effusion.
      d. *Increased teardrop* distance is more reliable evidence of hip effusion.
      e. *Hip aspiration* may require injection of 5 to 10 ml nonbacteriostatic saline and reaspiration. A small amount of contrast material confirms needle position in joint.
      f. *Differential diagnosis* of septic arthritis in the juvenile hip is extensive, including *transient synovitis, hemarthrosis* due to trauma or hemophilia, *JRA* and *early Legg-Calvé-Perthe's disease.*
   2. *Arthroplasties.*
      a. Infection is difficult to differentiate from simple loosening. Lucency at the

prosthesis or cement-bone interface is seen in either case.

b. *Periosteal reaction* and *heterotopic bone* formation are more suggestive of an infected arthroplasty.

c. Aspiration with culture should be a part of all arthrograms of prosthetic joints.

3. *Sacroiliac (SI) joint* septic arthritis.

a. Radiographic feature is *widening* of the SI joint with *indistinct cortices*. (Note that wide SI joints are normal in adolescents.)

b. Overlying bowel gas and the overlapped configuration of the joint may make plain film diagnosis difficult.

c. *CT* may be particularly useful since it demonstrates the widened joint, cortical erosions, and a soft tissue mass or psoas abscess. The differentiation of nonseptic arthritides or other causes of unilateral SI joint abnormalities is much easier with CT images.

4. *Sternoclavicular* septic arthritis.

a. The major differential diagnosis of a mass at the sternoclavicular junction is dislocation and septic arthritis (often in drug users).

b. *CT* limited to a few axial images through the joint will very clearly define erosion involving both sides of the joint and a soft tissue mass, easily differentiating septic arthritis from dislocation.

I. *Infection of the spine.*

1. Pertinent anatomy: *Veins of Batson* provide communication between the pelvic and thoracolumbar venous systems; thus, genitourinary (GU) infections are a frequent source of spine infection.

2. Despite the frequent GU source, *Staphylococcus aureus* is still the most common infecting organism.

3. Radiographic appearance:
   a. The infection starts in the subchondral portion of the body, then extends to the disk and adjacent vertebral end-plate.
   b. Radiographic changes are relatively late (2 to 8 weeks).
   c. Pathognomonic appearance therefore is two *adjacent end-plate abnormalities* (irregularity and loss of cortex) *with decrease in height of the corresponding disk.* A paravertebral mass or displaced psoas may be seen, especially by CT. Later, a sclerotic host reaction may evolve.
4. *TB of the spine (Pott's disease).*
   a. Usually there is pulmonary TB.
   b. Tends to involve the *thoracolumbar junction* and, late in the disease, may cause an acute *angular kyphosis* (gibbus).
   c. Features that favor the diagnosis include *slow progression,* with *preservation of disk height* and lack of sclerotic response.
   d. May develop a *calcified psoas abscess.*
5. *Discitis.*
   a. A less significant, often *self-limited,* disk infection in *skeletally immature* patients.
   b. Probably a direct hematogenous infection of the disk.
   c. Present with back pain, low-grade fever, elevated sedimentation rate; often there is a preexisting minor infection (e.g., upper respiratory).
   d. Radiographic changes are delayed but include *decrease in disk height, end-plate irregularities,* and *eburnation.* Paravertebral *soft tissue mass is minimal.*
   e. Bone scan demonstrates abnormality earlier than plain film.
   f. *Organisms are often not cultured,* either from blood culture or percutaneous biopsy. When proven, it is usually *Staphylococcus*

aureus. At some institutions, these patients are treated empirically for *Staphylococcus* infection.

6. *Discogenic sclerosis* (idiopathic segmental sclerosis).

   a. *A degenerative change that may mimic a disk space infection.*

   b. Degenerative disk leads to *decreased disk height.* Occasionally disk height is not decreased. *Adjacent vertebral end-plates are smooth* (except in the presence of a Schmorl's node—intravertebral disk herniation) *but sclerotic;* end-plate *sclerosis is often triangular in shape and may be limited to the anterior aspect of the upper involved vertebral body.*

   c. No progression (and even occasional resolution) with time.

III. SARCOIDOSIS

> **Key Concepts:** Lytic lesions with lacy trabeculae (usually in phalanges) or focal sclerotic lesions. Generally, if bone lesions are present, lung and skin abnormalities are as well.

A. Definition: A systemic granulomatous disorder; osseous involvement 1% to 15%.[4] If bone lesions are present, skin lesions usually are as well (90%). Lung abnormalities (hilar adenopathy, pulmonary infiltrates, fibrosis, and apical bullous disease) also usually are present (80% to 90%). Nodular liver disease with hepatosplenomegaly may be present, as may ocular abnormalities (uveitis, iritis).

B. Epidemiology:
   1. Young adults.
   2. No sex predominance.
   3. Blacks affected much more frequently than

Caucasians or Asians.
C. Osseous manifestations: There are several very different appearances.
   1. Generalized osteopenia.
   2. Lytic lesions with lacy trabeculae, usually in middle or distal phalanges.
   3. Sclerosis of tufts may occur.
   4. Focal or generalized sclerosis.
   5. Polyarticular arthralgias, usually without radiographic abnormalities but occasionally with nonspecific erosive arthritic changes.
D. Differential diagnoses:
   1. For lytic phalangeal lesions:
      a. Enchondroma.
      b. Brown tumor.
      c. Tuberous sclerosis.
   2. For lytic or sclerotic lesions at other sites:
      a. Paget's disease.
      b. Metastases.
      c. Myelofibrosis.
      d. Mastocytosis.

IV. RADIATION-INDUCED ABNORMALITIES*

A. Growth disorders: Depends on site of radiation.
   1. Epiphyseal fusion and limb length discrepancy.
   2. Scoliosis (if entire vertebral body is not included in the field).
   3. Iliac wing hypoplasia.
B. Radiation osteonecrosis.
   1. May appear aggressive (simulating tumor).
   2. Weakens the bone, allowing pathologic fractures.
   3. Avascular necrosis.
C. Neoplasm.
   1. Exostosis.
   2. Leukemia.
   3. Degeneration to osteosarcoma,

* See also Introduction to Chapter 1.

chondrosarcoma, or malignant fibrous histiocytoma.

D. More prone to infection than normal bone.

V. ACRO-OSTEOLYSIS

A. Definition: Lysis of the distal aspects of the phalanges, with a wide range of etiologies.
B. Gamut.
   1. Thermal injury.
      a. Burn: May have associated soft tissue contracture.
      b. Frostbite: Characteristically, the thumb is spared. In children, the distal epiphyses are most at risk and may either resorb or become sclerotic.
   2. Environmental: Polyvinylchloride (PVC) exposure causes acro-osteolysis. The most characteristic pattern is a lucent transverse zone across the proximal to middle aspect of the distal phalanges.
   3. Metabolic:
      a. Hyperparathyroidism: Tuft resorption, often accompanied by other signs such as subperiosteal resorption, vascular calcification, or brown tumors.
      b. Lesch-Nyhan disease.
   4. Arthritis:
      a. Psoriatic: There should be associated DIP and carpal erosive disease.
      b. Neuroarthropathy, especially diabetic.
   5. Connective tissue disease:
      a. Scleroderma: Often associated soft tissue calcification.
      b. Other causes of vasculitis.
   6. Infection: Leprosy: May see associated linear calcification of a digital nerve.
   7. Congenital:
      a. Pyknodysostosis: Associated dense bones with transverse fractures.

b. Hajdu-Cheney: Familial disorder with
acro-osteolysis pattern identical to that of
PVC, associated with osteoporosis and
multiple facial and cranial abnormalities.

VI. PERIOSTEAL REACTION IN INFANTS

> **Key Concepts:**   *Age at onset may be useful in diag-nosis;* must remember to consider nonaccidental trauma (battered child syndrome) as a common cause.

A. *Onset prior to age 6 months:*
1. *Infantile cortical hyperostosis* (Caffey's disease).
   a. Onset early, occasionally at birth.
   b. Clinical findings: Fever, *hyperirritability,* and soft tissue swelling (reflects the periosteal reaction of underlying bone).
   c. Usually self-limited but may persist, leading to a delay in musculoskeletal development.
   d. X-ray findings:
      (1) *Involves mandible, clavicles,* ribs, scapulae, cranium, and tubular bones, often sequentially.
      (2) The mandibular and clavicular involvement are important findings since they are generally not present in infant periosteal reaction of other etiologies.
      (3) Cortical hyperostosis may be marked, sometimes with osseous bridging.
      (4) No metaphyseal abnormalities.
2. *Physiologic periosteal new bone.*
   a. Occurs in 35% of infants after 1 month of age.
   b. Bilateral, in long tubular bones.
   c. Incorporated into cortex by age 6 months.
   d. Diagnosis of exclusion.

B. *Onset after age 6 months:*
  1. *Hypervitaminosis A.*
     a. Cortical thickening of tubular bones, most commonly the ulna; wavy or undulating pattern.
     b. Metaphyseal cupping, most commonly distal femurs, may result in coned epiphyses.
     c. Findings of increased intracranial pressure.
     d. Child may be extremely irritable.
  2. *Scurvy or rickets* may show periosteal reaction secondary to a metaphyseal corner fracture and subperiosteal bleed. Prior to 6 months they occur only in the severely stressed premature infant.
C. *Onset at any time during infancy:*
  1. Trauma: *Nonaccidental trauma* is a common occurrence and may be radiographically manifest solely by periosteal reaction (for a more complete description of this entity, see the section of the handbook on trauma).
  2. *Infection:* Many pathogens may be involved, but congenital syphilis should especially be considered.
     a. Osteochondritis with widened zone of provisional calcification and metaphyseal irregularity in tubular bones and costochondral junctions seen in fetus and neonate.
     b. Diaphyseal osteomyelitis: A later change, seen with inadequate therapy.
     c. Periostitis: Widespread and symmetric.
  3. Periosteal response to *tumor.*
     a. Neuroblastoma metastasis: Generally metaphyseal permeative change, but periosteal reaction may be the first manifestation.
     b. Periostitis in acute childhood *leukemia.*

      (1) Leukemia may have a paucity of
clinical signs.

      (2) Arthralgias and arthritis are common
(12% to 65%) and may be the
presenting symptom.

      (3) Periostitis seen in 10% to 35%, due to
invasion of subperiosteum by tumor,
elevating the periosteum.

      (4) Especially prominent in the terminal
phalanges.

4. Prostaglandin administration may cause
periosteal reaction.

5. Sickle cell dactylitis (infarcts, generally in the
phalanges, with associated periosteal
reaction).

6. In patients past infancy JRA should be
considered since periostitis, especially of the
phalanges, is a common feature of that
disease process.

7. "Wavy" radius is a normal variant that
simulates periosteal reaction.

VII. LOCALIZED GIANTISM

A. Definition: Overgrowth of both osseous and soft
tissues localized to a few digits.

B. *Neurofibromatosis* may have an associated
dysplasia causing localized giantism.

C. *Hypervascularity* (as from an arteriovenous
malformation) is a second common cause of
localized giantism; another is
lymphangiomatosis.

D. *Macrodystrophia lipomatosa* is a rare cause of
localized giantism of unknown etiology; it is
identified pathologically by a disproportionate
increase in adipose tissue and osteoblasts lining
the periosteum. Second and third digits of the
lower extremity are the most common sites.

E. *Klippel-Trenaunay-Weber syndrome* is another rare
etiology of localized giantism. It is associated

with cutaneous capillary hemangioma and varicose veins. Giantism is attributed to the abnormal vascular supply. Phleboliths may be present.

VIII. SOFT TISSUE CALCIFICATION GAMUT

> **Key Concepts:** Presumes normal underlying bone. Excludes vascular calcification and chondrocalcinosis. Details of each entity are discussed in separate chapters.

A. Trauma:
  1. Myositis ossificans: Has a characteristic zoning phenomenon and timing of appearance and maturation.
  2. Burns: Contractures often are associated.
  3. Frostbite.
  4. Head injury.
  5. Paraplegia or quadriplegia.
  6. Hydroxyapatite crystal deposition disease, especially calcific bursitis or tendinitis.
B. Tumor: Any soft tissue tumor may have dystrophic calcification but those most commonly considered include:
  1. Synovial cell sarcoma.
  2. Liposarcoma.
  3. Soft tissue osteosarcoma.
  4. Fibrosarcoma or malignant fibrous histiocytoma.
C. Collagen vascular diseases:
  1. Scleroderma: Calcification is usually subcutaneous, may be widespread; usually other changes of scleroderma are present.
  2. Dermatomyositis: Sheetlike calcification in muscle or fascial planes are described, but other calcification patterns are seen as well.
  3. Systemic lupus erythematosus: Calcification is uncommon but certainly may occur, even in

the absence of renal disease.
4. CRST syndrome—calcinosis cutis, Raynaud's, scleroderma, and telangiectasis.
5. Calcinosis cutis.
D. Congenital:
   1. Tumoral calcinosis: Periarticular.
   2. Myositis ossificans progressiva: Often axial, bridging bones of the thorax.
   3. Pseudohypoparathyroidism and pseudopseudohypoparathyroidism.
   4. Progeria.
   5. Ehlers-Danlos syndrome.
E. Metabolic:
   1. Hyperparathyroidism, primary or secondary.
   2. Hypoparathyroidism.
   3. Gout.
   4. Renal dialysis sequela: Periarticular calcifications may be very extensive.
F. Infections:
   1. Granulomatous: TB, brucellosis, coccidioidomycosis.
   2. Dystrophic calcification in abscesses.
   3. Leprosy: Digital nerve calcification.
   4. Cysticercosis: Small oval bodies in muscle.
   5. Echinococcosis: Usually in liver or bone but occasionally in soft tissues.
G. Drugs:
   1. Hypervitaminosis D.
   2. Milk-alkali syndrome.
H. Miscellaneous conditions:
   1. Synovial chondromatosis.
   2. Weber-Christian disease.

REFERENCES

1. Al-Sheikh W, Sfakionakis G, Mnaymneh W, et al.: Subacute and chronic bone infections: Diagnosis using In-111, Ga-67, and Tc-99m bone scintigraphy, and radiography. *Radiology* 1985; 155:501–506.

2. Alazraki N, Fierer J, Resnick D: Chronic osteomyelitis monitoring by Tc99m phosphate and Ga 67 Citrate imaging. *Am J Roentgenol* 1985; 145:767–771.
3. Traughber P, Manaster B, Murphy K, et al.: Negative bone scans of joints after aspiration and/or contrast arthrography: Experimental studies. *Am J Roentgenol* 1986; 146:87–92.
4. Sartoris D, Resnick D, Resnick C, et al.: Musculoskeletal manifestations of sarcoidosis. *Semin Roentgenol* 1985; 20:376–386.

Index

A

Abscess
  Brodie's, 385–386
    differential diagnosis,
      22
  psoas, calcified, in tuber-
    culosis of spine, 392
Absorptiometry: dual-pho-
  ton, in osteoporosis,
  297
Abuse: drug, and infection,
  382
AC
  joints, 226–227
  separation, 226–227
Acetabular
  dysplasia in hip disloca-
    tion, congenital, 334
  roofs, horizontal, in
    achondroplasia, 357
Acetabuloplasty: Pember-
    ton, in hip disloca-
    tion, congenital, 335

Acetabulum, primary pro-
    trusio of, 337–338
  definition, 337
  diagnosis, differential,
    337–338
Achondrogenesis, 361
  clinical features, 361
  diagnosis, differential,
    361
  inheritance, 361
  prognosis, 361
  progression, 361
  radiographic features, 361
  type I, 361
  type II, 361
Achondroplasia, 356–358
  clinical features, 357
  diagnosis, differential,
    357–358
  inheritance, 356–357
  prognosis, 357
  progression, 357

Achondroplasia (*cont.*)
  radiographic features, 357
  synonyms, 356
Acromegaly, 295, 311–312
  definition, 311
  foot in, 311–312
  growth hormone in, 311
  hand in, 311–312
  radiographic abnormalities, 311–312
Acromelia, 372
Acro-osteolysis, 319, 395–396
  arthritis in, 395
  congenital, 395–396
  connective tissue disease and, 395
  definition, 395
  gamut, 395–396
  infection in, 395
  metabolic factors in, 395
  in sclerosing dysplasia, 332
Acropachy, thyroid, 311
  differential diagnosis, 182
  periosteal reaction in, 311
Adamantinoma, 88–90
  differential diagnosis, 25
  major, 89
  host response, 89
  key concepts, 88
  radiographic work-up, 89–90
  treatment, 90
Age
  myeloma and, multiple, 68
  tumors and, 2–3
Aged (*see* Osteosarcoma, in aged)
Air arthrography: in septic arthritis of hip in children, 390

Albers-Schonberg disease: in sclerosing dysplasia, 331
Alcohol: bone abnormalities due to, 317
Alcoholism, 295
  avascular necrosis and, 183
Alkaline phosphatase, 318
  in osteogenesis imperfecta, 330
  in Paget's disease of bone, 315
Alkaptonuria, 174–175
  definition, 174
  differential diagnosis, 175
  epidemiology, 174
  joints commonly affected, 174
  key concepts, 174
  radiography of, 174
  signs, clinical, 174
  spine in, 174
Aluminum intoxication: in renal osteodystrophy, 307
Amyloidosis, 145–147
  definition, 146
  differential diagnosis, 147
  epidemiology, 146
  extra-articular manifestations, 146
  joints commonly affected, 147
  key concepts, 145
  laboratory tests, 146
  radiography in, 146–147
  signs, clinical, 146
Anemia, 314, 380–382
  Cooley's, 381–382
  definition, 380
  features, 381–382
  infarction in, 381
  key concepts, 380
  radiographic abnormalities, 380–381

Anemia (*cont.*)
  sickle cell, 381
Aneurysm: shunt, in renal
    osteodystrophy, 308
Aneurysmal (*see* Cyst,
    bone, aneurysmal)
Angioblastoma (*see*
    Adamantinoma)
Angiography
  in tumor work-up, 13
  of vascular metastases,
    103
Angiomatosis, cystic, 73–74
  differential diagnosis,
    major, 73
  key concepts, 73
  radiographic work-up, 74
  treatment, 74
Angiosarcoma, 76–78
  differential diagnosis,
    major, 77
  key concepts, 77
  radiographic work-up,
    77–78
  treatment, 78
Ankle
  in arthritis, juvenile
    chronic, 122
  effusion, 256
  fracture (*see* Fracture,
    ankle)
  in gout, 163
  in hemarthroses, 379
  normal, AP view, 255
  postoperative evaluation,
    257
  trauma, 254–260
    key concepts, 254
    Lauge-Hansen classifi-
      cation, 256
    structures to be exam-
      ined, 256
    Weber classification,
      256–257

Ankylosing (*see* Spondylitis,
    ankylosing)
Ankylosis
  in arthritis
    juvenile chronic, 122
    rheumatoid, 112
  in spondylitis, 126
Anomalies
  congenital, 320–377
    of cervical spine, 269–
      273
  foot (*see* Foot, anomalies)
Aponeurotic fibroma: juve-
    nile, 78
Apophyseal avulsion, 237–
    238
Apposition of fracture (*see*
    Fracture, apposition
    of)
Arachnodactyly
  in Ehlers-Danlos syn-
    drome, 329
  in homocystinuria, 328
  in Marfan's syndrome,
    327
Arthritis, 107–188
  in acro-osteolysis, 395
  generalizations, 107–110
  juvenile chronic
    ankle in, 122
    differential diagnosis,
      124
    elbow in, 122
    feet in, 122
    hand in, 122
    hip in, 123
    joints commonly af-
      fected, 122
    knee in, 122–123
    radiography of, survey
      films for early diag-
      nosis, 124
    shoulder, in, 122
    spine in, 123

Arthritis (*cont.*)
  symmetry in, 123
  temporomandibular
    joint in, 123
  wrist in, 122
osteoarthritis (*see*
  Osteoarthritis)
polyarthritis of unknown
  etiology, 110–138
psoriatic, 130–135
  definition, 130
  differential diagnosis,
    134
  epidemiology, 130
  extra-articular manifes-
    tations, 131
  feet in, 132
  hand in, 132
  joints most commonly
    affected, 132–134
  key concepts, 130
  laboratory tests in, 131
  radiography of, 131–
    132
  radiography of, survey
    films for early diag-
    nosis, 134–135
  signs, clinical, 131
  spine in, 133–134
  symmetry in, 134
rheumatoid (*see* Rheuma-
  toid arthritis)
septic, 389–391
  arthroplasty in, 390–
    391
  diagnosis, special con-
    siderations in, 390–
    391
  fungal, 390
  joints in, 389
  radiographic signs, 389
  renal osteodystrophy
    and, 308
  sacroiliac joint in, 391

sternoclavicular, 391
Arthrography
  air, in septic arthritis of
    hip in children, 390
  in hip dislocation, con-
    genital, 335
  knee, 253–254
  tarsal coalition, 353–354
  wrist, 217, 218
Arthrogryposis multiplex
  congenita, 324–325
  clinical features, 324
  distribution, 325
  etiology, 324
  key concepts, 324
  radiographic findings,
    324–325
Arthropathy
  calcium pyrophosphate,
    167
  diabetic, 160
  hemochromatosis (*see*
    Hemochromatosis
    arthropathy)
  in hemophilia, 379
  inflammatory bowel dis-
    ease, 121
  Jaccoud's, 148
  neuroarthropathy (*see*
    Neuroarthropathy)
  osteoarthropathy (*see*
    Osteoarthropathy)
  psoriatic, 121
  pyrophosphate, 165
Arthroplasty
  in septic arthritis, 390–
    391
  Swanson, 291
Arthroplasty, total hip,
  285–288
  bone imaging, 287–288
  endoprosthesis in, 288
  evaluation
    for lengths of gluteus
      medius and iliopsoas
      muscle groups, 286

Arthroplasty, total hip
(*cont.*)
for loosening, 286
of placement, 285
infection, 287
loosening pattern, 287
resorption around a fem-
oral component, 286
revision arthroplasty, 288
sclerosis around femoral
component, 287
Arthroplasty, total knee,
288–291
carpal implants in, 291
cerclage wires in, 290
evaluation of loosening,
289
femur in, 289
fixators in, external, 290
intermedullary rod in,
290
normal placement, 288
Parham bands in, 290
patellar complications,
289
screws in, dynamic, 290
spinal instrumentation
in, 291
Arthroscopy: of knee, 253
Aseptic necrosis, 295
Asphyxiating (*see* Dysplasia, asphyxiating
thoracic)

Atlantoaxial
distance, 267
impaction in rheumatoid
arthritis, 117
instability in Morquio's
syndrome, 376
rotary displacement, 273
subluxation in rheuma-
toid arthritis, 117
radiography of, 118
Atlas, 273
occipitalization of, 269–
270
Atrophic
neuropathic joints, 159
nonunion, 194
Atrophy: Sudeck's, 296
Avascular (*see* Necrosis,
avascular)
AVN: in anemia, 381
Avulsion, 193
apophyseal, 237–238
cruciate ligament, 250
epicondyle, medial, 223
of flexor digitorum pro-
fundus, 199
fracture, 196–200
of lesser trochanter,
244
phalangeal, sites of,
196
of volar fragment of pha-
lanx, 199–200

B

"Bamboo" spine, 128
Barton's fracture, 206
reverse, 206
Baseball finger, 197
Battered child syndrome,
279–281
bone imaging in, 280
epidemiologic considera-
tions, 280–281
key concepts, 279

radiographic signs con-
sidered to be patho-
gnomonic, 279–280
Bennett variety: of fracture
of first metacarpal,
201
Biochemical abnormalities,
161–175
Biopsy: in tumor work-up,
10

Bismuth: bone abnormalities due to, 317
Blount's disease, 179
Blue sclerae: in osteogenesis imperfecta, 330
Bone
  abnormalities
    drug-induced, 317
    environmentally-induced, 319
  cyst (*see* Cyst, bone)
  destruction in massive osteolysis, 74
    pattern in tumors, 3
  disease
    metabolic, 293–319
    metabolic, laboratory values, serum, 319
    in renal osteodystrophy, 307
  erosion in septic arthritis, 389
  facial (*see* Facial bones)
  -forming tumors, 17–41
    benign, 15, 17–27
    malignant, 15, 27–40
    WHO classification, with modification, 15
  heterotopic, in septic arthritis, 391
  imaging, 102
    in arthroplasty, total hip, 287–288
    in battered child syndrome, 280
    in tarsal coalition, 352
  island (*see* Enostosis)
  long
    in achondrogenesis, 361
    in chondroectodermal dysplasia, 363
    mucopolysaccharidoses, 376

  in osteogenesis imperfecta, 330
  in sclerosing dysplasia, 331
  in scurvy, 312
  in spondyloepiphyseal dysplasia congenita, 368
  lymphoma (*see* Lymphoma, bone)
  mass quantitative analysis in osteoporosis, 297
  metastases (*see* Metastases, bone)
  Paget's disease (*see* Paget's disease of bone)
  physiologic periosteal new, in periosteal reaction in infant, 396
  resorption in scleroderma, 143
  sesamoid, 262
  in spondylitis, 125
  subperiosteal, in scurvy, 312
  tubular
    in achondroplasia, 357
    in anemia, 380
    in chondrodysplasia calcificans punctata, 365
    in osteomyelitis, hematogenous spread, 384
    in Paget's disease of bone, 316
    rhizomelic, in thanatophoric dwarfism, 360
  tumors
    bone-forming tumors (*see above*)
    characteristics, 1
    staging, surgical, 6
    -within-a-bone in sclerosing dysplasia, 331

Bone (*cont.*)
  wormian (*see* Wormian bone)
Bouchard nodes, 150
Boutonnière deformity, 197–199
  lateral view of, 198
Bowel: inflammatory bowel disease, 121, 124–130
Bowman's angle, 219
Boxer's fracture, 200
Brodie's abscess, 385–386
  differential diagnosis, 22

Bronze skin: with cirrhosis and diabetes, 169
Brown tumor, 101
  differential diagnosis, 24, 60
  in hyperparathyroidism, 306
Burns, 319
  in acro-osteolysis, 395
  with myositis ossificans, 283
Bursitis: retrocalcaneal, in Reiter's disease, 136
Butterfly fracture, 192

C

Caffey's disease, 396
Caisson disease: and avascular necrosis, 183
Calcaneonavicular coalition, 350
  medial oblique view, 351
Calcaneus, 260, 338
  hindfoot, 339, 340
  trauma, 262
Calcification
  in alkaptonuria, 174
  in Ehlers-Danlos syndrome, 329
  in hyperparathyroidism, 306
  in hypoparathyroidism, 309
  in osteoarthritis, 150
  in osteodystrophy, renal, 308
  in pseudohypoparathyroidism, 309–310
  in scleroderma, 142
  in scurvy, 312
  soft tissue (*see* Soft tissue, calcification)
  in synovial cell sarcoma, 86

  in tuberculosis of spine, 392
Calcinosis: tumoral, differential diagnosis, 283–284
Calcium, 318
  deprivation in osteomalacia and rickets etiologies, 300
Calcium hydroxyapatite deposition disease, 172–173
  definition, 172
  differential diagnosis, 173
  elbow in, 173
  epidemiology, 172
  hand in, 173
  joints commonly affected, 173
  key concepts, 172
  laboratory tests, 172
  radiography of, 172
  shoulder in, 173
  wrist in, 173
Calcium pyrophosphate dihydrate crystal deposition disease, 164–168
  definition, 164–165

Calcium pyrophosphate
        dihydrate crystal
        deposition disease
        (*cont.*)
    differential diagnosis, 168
    elbow in, 168
    epidemiology, 165
    hand in, 168
    hip in, 168
    joints commonly affected,
        166–168
    key concepts, 164
    knee in, 166–167
    laboratory tests, 165–166
    radiography of, 166
        survey films, 168
    shoulder in, 168
    signs, clinical, 165
    wrist in, 167
Callus formation: in osteo-
        genesis imperfecta,
        330
Calvaria: in achondroplasia,
        357
Capitate
    dorsiflexion of, 216
    fracture, 211
    fusion in chondroectoder-
        mal dysplasia, 363
    -lumate instability, 214
Capsular sign: lateral, 250
Cardiomegaly: in anemia,
        381
Carpal tunnel syndrome: in
        renal osteodystro-
        phy, 308
Carpometacarpal joints: ar-
        ticulation, 202
Carpus
    dislocation, 211–214
        patterns seen on lateral
            film, 213
    in dyschondrosteosis, 367
    fracture, 210–211

fracture dislocation, 212
fusions, 208–209
implant in total knee ar-
        throplasty, 291
instabilities, 214–216
    dorsiflexion, 216
    fluoroscopy of, 214
    patterns of, 215
    volar flexion, 216
normal anatomy, 207–209
    carpal tunnel view, 209
    lateral view, 209
normal lateral view, 209
normal PA view, 207
normal variants, 208
subluxation, dorsal, 216
trauma, 206–216
    frequency of, 206–207
Cartilage
    in acromegaly, 312
    destruction
        in arthritis, juvenile
            chronic, 122
        in gout, 162
        in osteoarthritis, 150
        in rheumatoid arthritis,
            111
    -forming tumors, 40–58
        benign, 40–52
        malignant, 52–58
        WHO classification
            with modification, 15
    in hemophilia, 379
    in septic arthritis, fungal,
        390
Cavus, 338
Cell(s)
    clear (*see* Chondrosar-
        coma, clear cell)
    giant (*see* Giant cell)
    mast, disorder in masto-
        cytosis, 312
    sickle (*see* Sickle cell)

Cerclage wires: in total
  knee arthroplasty,
  290
Cervical
  lordosis, 264
  spine (*see* Spine, cervical)
Chance fracture, 277–278
Charcot joint (*see* Neuropathic joints)
Charcot-like joints, 295
Chauffeur's fracture, 206
Chemotherapy
  complications of, 14
  of osteosarcoma, 31–32
Chiari medial displacement:
  in hip dislocation,
  congenital, 336
Children
  battered (*see* Battered
    child syndrome)
  bone metastases, differ-
    ential diagnosis, 105
  fractures, 194–195
    elbow, 222–223
  humerus separation in,
    224
  hyperthyroidism in, 311
  metaphyseal fracture,
    transverse, 205
  osteomyelitis, hematoge-
    nous spread, 384
  scurvy, 312
  septic arthritis, hip in,
    390
Cholecalciferol (*see* Vitamin
  D)
Cholelithiasis: in anemia,
  381
Chondroblastoma, 47–49
  age and, 3
  differential diagnosis, 60
    major, 49
  key concepts, 47
  radiographic work-up, 49

of tibia, 48
  treatment, 49
Chondrocalcinosis, 164
  in CPPD crystal deposi-
    tion disease, 166
  in hyperparathyroidism,
    306
  in menisci, 166
  in osteoarthritis, 150
Chondrodysplasia calcifi-
    cans punctata, 364–
    366
  clinical features, 365
  diagnosis, differential,
    366
  inheritance, 365
  prognosis, 366
  progression, 366
  radiographic features, 365
  synonyms, 364–365
Chondrodysplasia: meta-
    physeal, Schmid
    type, mimicking rick-
    ets, 302
Chondroectodermal (*see*
    Dysplasia,
    chondroectodermal)
Chondroma, 40–44
  differential diagnosis,
    major, 41–42
  juxtacortical, 50–52
    differential diagnosis,
      283
    differential diagnosis,
      major, 51
    key concepts, 51
    radiographic work-up,
      52
    treatment, 52
  key concepts, 40
  osteochondroma (*see*
    Osteochondroma)
  periosteal, 50–52
  radiographic work-up, 42

Chondroma (*cont.*)
  treatment, 42–43
Chondromatosis, synovial,
    176–177
  definition, 176
  epidemiology, 176
  joints commonly affected,
    177
  key concepts, 176
  radiography of, 176–177
  signs, clinical, 176
Chondromyxoid fibroma
    (*see* Fibroma,
    chondromyxoid)
Chondrosarcoma, 52–58
  age and, 3
  central, 52–55
    differential diagnosis,
      major, 54
    in iliac wing, 53
    radiographic work-up,
      55
    survival, 55
    treatment, 55
  clear cell, 57
    differential diagnosis,
      major, 57
    radiographic work-up,
      57
    treatment, 57
  dedifferentiated, 39, 57–
    58
    prognosis, 58
    treatment, 58
  differential diagnosis, 45
  extraskeletal, 86
  key concepts, 52
  medullary (*see* central
    above)
  mesenchymal, 58
    radiography, 58
    treatment, 58
  peripheral, 55–56
    differential diagnosis,
      major, 56

    radiographic work-up,
      56
    treatment, 56
Chordoma, 86–88
  differential diagnosis,
    major, 87–88
  key concepts, 87
  radiographic work-up, 88
  treatment, 88
Christmas disease, 378
Chromosome X deletion,
    355
Cirrhosis: with bronze skin
    and diabetes, 169
Clavicle, 226
  absence in cleidocranial
    dysostosis, 354
  fracture, 226
  in periosteal reaction in
    infant, 396
  resorption, differential di-
    agnosis, 226
Clay-shoveler's fracture,
    275
Cleidocranial (*see* Dysosto-
    sis, cleidocranial)
Closed fracture, 191
Clubbed fingers: in thyroid
    acropachy, 311
Clubfoot, 343–346
  in dwarfism, diastrophic,
    364
  incidence, 343
  radiographic findings,
    345–346
Collagen
  in scurvy, 312
  synthesis
    defect in homocystin-
      uria, 328
    diseases of, 322
  vascular disease and soft
    tissue calcification,
    399–400

Colles' fracture, 205–206
  reverse, 206
Comminution fracture, 192
Complete fracture, 191–192
Computed (*see* Tomography, computed)
Condylar fracture: lateral, 223
Congenital anomalies, 320–377
Connective tissue disorders, 139–147
  in acro-osteolysis, 395
  Marfan's syndrome as, 327
Connective tissue tumors, 78–86
  benign, 78–82
Contractures, 319
  flexion
    in hemarthroses, 378
    in homocystinuria, 328
  joint, in diastrophic dwarfism, 364
Cooley's anemia, 381–382
Copper deficiency: in prematurity, 302
Cortical defect, benign fibrous (*see* Fibroma, nonossifying)
Coumadin: causing bone abnormalities, 317
Coxa vara, infantile, 337
  definition, 337
  diagnosis, differential, 337
Cranium
  in cleidocranial dysostosis, 354

Dactylitis
  in anemia, 380, 381

in Paget's disease of bone, 316
Crescent sign, 184
Cruciate ligament: avulsion, 250
Cushing's disease: differential diagnosis, 295–296
Cyst
  bone, aneurysmal, 92–95
    age and, 3
    differential diagnosis, major, 94
    key concepts, 94
    radiographic work-up, 95
    treatment, 95
  bone, solitary, 90–92
    differential diagnosis, major, 92
    key concepts, 90
    radiographic work-up, 92
    treatment, 92
  "dialysis," 308
  subarticular, in hemophilia, 379
  subchondral
    in CPPD crystal deposition disease, 166
    gout and, 163
    in osteoarthritis, 151
    in rheumatoid arthritis, 112
Cystathionine synthetase: in homocystinuria, 328
Cystic (*see* Angiomatosis, cystic)

D

  sickle cell, and periosteal reaction in infant, 398
Dentition: hypoplastic, in hypoparathyroidism, 309

Dermal fibromatosis: infantile, 78
Dermatomyositis (*see* Polymyositis/dermatomyositis)
Desmoid
cortical, 80–81
differential diagnosis, 29
periosteal, 80–81
tumor, 79
Desmoplastic (*see* Fibroma, desmoplastic)
Diabetes mellitus
arthropathy in, 160
with bronze skin and cirrhosis, 169
osteomyelitis of foot in, 388–389
Dialysis
abnormalities after, and renal osteodystrophy, 308–309
"cysts," 308
Diaphyseal infarcts, 295
in anemia, 380–381
Diaphysis: tumors in, 4
Diastasis, 193
Diastrophic (*see* Dwarfism, diastrophic)
Didronyl: osteomalacia and rickets, 301
Dilantin
bone abnormalities due to, 317
osteomalacia and rickets, 301
Disc: degenerative disc disease, 156
Discitis, 392–393
radiographic changes, 392
in rheumatoid arthritis, 119

Discogenic sclerosis, 393
Dish (*see* Hyperostosis, diffuse idiopathic skeletal)
Dislocation, 193
carpus, 211–214
pattern seen on lateral film, 213
in Ehlers-Danlos syndrome, 329
elbow, 223–224
congenital, 224
myositis ossificans complicating, 224
fibula, proximal, 250
fracture (*see* Fracture-dislocation)
hand, 201–203
hip, 245–246
hip, congenital, 332–336
arthography in, 335
complications, 335
CT of, 355
definition, 333
epidemiology, 333
films before 6 weeks have high false-negative rate, 333
key concepts, 332
radiographic signs, 333–335
treatment 335–336
ultrasound in, 335
knee, 252–253
lunate, 213–214
metacarpal, 201–203
midcarpal, 212–213
patella, 251–252
perilunate, 212
phalanx, 201
radial head, 224
radioulnar, 205

Dislocation (*cont.*)
  shoulder, 229–230
    anterior, 229–230
    inferior, 230
    posterior, 230
  talonavicular joint, 348
  ulna in dyschondrosteo-
    sis, 367
Down's syndrome, 335
Drug(s)
  abuse, and infection, 382
  -induced bone abnormali-
    ties, 317
  -related osteoporosis, 296
  soft tissue calcification
    and, 400
Dwarfism, 356–373
  diastrophic, 364
    clinical features, 364
    diagnosis, differential,
      364
    inheritance, 364
    prognosis, 364
    progression, 364
    radiographic features,
      364
  flow chart for characteri-
    zation of, 369
  key concepts, 356
  radiographic classifica-
    tion, approach to,
    368–373
  thanatophoric, 359–361
    clinical features, 360
    diagnosis, differential,
      360
    inheritance, 360
    prognosis, 360
    progression, 360
    radiographic features,
      360
    synonyms, 359
Dyschondrosteosis, 366–367
  clinical features, 366

  diagnosis, differential,
    367
  inheritance, 366
  prognosis, 367
  progression, 367
  radiographic features,
    366–367
  synonyms, 366
Dysostosis
  cleidocranial
    definition, 354
    radiographic abnormal-
      ities, 354
  osteo-onychodysostosis
    (*see* Osteo-
    onychodysostosis)
Dysplasia
  acetabular, in hip disloca-
    tion, congenital, 334
  asphyxiating thoracic,
    361-362
    clinical features, 361–
      362
    diagnosis, differential,
      362
    prognosis, 362
    progression, 362
    radiographic features,
      362
    synonyms, 361
  chondrodysplasia (*see*
    Chondrodysplasia)
  chondroectodermal, 362–
    363
    clinical features, 362–
      363
    diagnosis, differential,
      363
    prognosis, 363
    progression, 363
    radiographic features,
      363
    synonyms, 362
  diastrophic (*see* Dwarf-
    ism, diastrophic)

Dysplasia (*cont.*)
  epiphysialis hemimelica,
    46–47
  fibrous, 99–101
    differential diagnosis,
      major, 101
    key concepts, 99
    radiographic work-up,
      101

Ehlers-Danlos syndrome,
    328–329
  definition, 329
  epidemiology, 329
  key concepts, 329
  musculoskeletal features,
    329
  nonmusculoskeletal fea-
    tures, 329
Elbow
  AP film, 218
  in arthritis juvenile
    chronic, 122
  calcium hydroxyapatite
    deposition disease,
    173
  in CPPD crystal deposi-
    tion disease, 168
  dislocation (*see* Disloca-
    tion, elbow)
  fracture (*see* Fracture,
    elbow)
  in hemarthroses, 379
  in hemophilia, 379
  lateral film, 218
  normal anatomy, 218–221
  normal variants, 221–222
  nursemaid's, 224
  in rheumatoid arthritis,
    113
  trauma, 218–224
    key concepts, 218
Embolization: of vascular
    metastases, 103

treatment, 101
  sclerosing, 331–332
    definition, 331
    key concepts, 331
  spondyloepiphyseal (*see*
    Spondyloepiphyseal
    dysplasia)
Dystrophy: reflex sympa-
    thetic, 296

E

Enchondroma (*see*
    Chondroma)
Enchondromatosis: multi-
    ple, 43
Endoprostheses: in total
    hip arthroplasty, 288
Endosteal resorption: in hy-
    perparathyroidism,
    305
Endosteal scalloping: in os-
    teomyelitis, 384
Enneking methodology, 6
Enostosis, 18–19
  differential diagnosis, 22
    major, 19
  key concepts, 18
  radiography in, 19
  in sclerosing dysplasia,
    332
  in hyperostosis, diffuse
    idiopathic skeletal,
    186
  in osteoarthritis, 151
Environmentally induced
    bone abnormalities,
    319
Eosinophilic granuloma:
    differential diagno-
    sis, 385
Epicondyle avulsion: me-
    dial, 223

Epiphysis
  in chondrodysplasia calci-
      ficans punctata, 365
  of extremities, in spondy-
      loepiphyseal dyspla-
      sia congenita, 368
  femoral (*see* Femur,
      epiphysis)
  hemophilia, 379
  in hypothyroidism, juve-
      nile, 310
  knee, injury, 250–251
  maturation sequence, 222
  normal, in achondropla-
      sia, 357
  in pseudoachondroplasia,
      359
  radial, Salter fracture of,
      205
  slipped, in renal osteo-
      dystrophy, 307
  spondyloepiphyseal (*see*
      Spondyloepiphyseal)
  stippling
      in chondrodysplasia
          calcificans punctata,
          365
      in juvenile hypothy-
          roidism, 310
  tibia, fusion, 258

Facet erosion: in rheuma-
      toid arthritis, 118
Facial bones
  in achondroplasia, 357
  in acromegaly, 311
Factor
  VIII, 378
  IX, 378
Feet (*see* Foot)
Felty's syndrome, 119
Femur
  in arthroplasty, total
      knee, 289

  tumors in, 4
Equinus, 338
  hindfoot (*see* Hindfoot,
      equinus), Relen-
      meyer flask
      deformity
  diagnosis, differential,
      314
  in Gaucher's disease, 313
Ewing's sarcoma (*see* Sar-
      coma, Ewing's)
Exostoses
  (*See also*
      Osteochondroma)
  in neurofibromatosis, 326
  in pseudohypoparathy-
      roidism, 310
Exostotic chondrosarcoma
  (*see* Chondrosar-
      coma, peripheral)
Extremities
  in asphyxiating thoracic
      dysplasia, 362
  epiphyses in spondylo-
      epiphyseal dysplasia
      congenita, 368
  lower, osteomyelitis in,
      hematogenous
      spread, 384
  salvage with allograft, 14

  in chondrodysplasia calci-
      ficans punctata, 365
  condyle fracture, 248
  epiphysis, capital
      in hypochondroplasia,
          358
      in hypothyroidism, ju-
          venile, 310
  epiphysis, slipped capi-
      tal, 246
      AP of, 247
      epidemiology, 246
      radiography of, 246

F

Femur (*cont.*)
  SCFE, 246
  focal deficiency, proximal, 336–337
    definition, 336
    radiographic abnormalities, 336–337
  fracture (*see* Fracture, femur)
  head
    anteversion in hip dislocation, congenital, 334–335
    ossification center in hip dislocation, congenital, 333
  neck, osteoid osteoma in, 20
  resorption around a component of in total hip arthroplasty, 286
  "telephone receiver-shaped," in thanatophoric dwarfism, 360
Fever (*see* Rheumatic fever)
Fibroma
  aponeurotic, juvenile, 78
  chondromyxoid, 49–50
    differential diagnosis, major, 50
    key concepts, 49
    radiographic work-up, 50
    treatment, 50
  desmoplastic, 79-80
    differential diagnosis, major, 80
    radiographic work-up, 80
    treatment, 80
  nonossifying, 95–96
    differential diagnosis, major, 96
    key concepts, 95

  radiographic work-up, 96
    treatment, 96
  ossifying, 25–27
    differential diagnosis, major, 25
    key concepts, 25
    radiographic work-up, 26
    of tibia, 26
    treatment, 26–27
Fibromatosis, 78–81
  aggressive, 79
  congenital generalized, 78–79
  dermal, infantile, 78
Fibrosarcoma, 82–84
  age and, 3
Fibrous cortical defect (*see* Fibroma, nonossifying)
Fibrous dysplasia (*see* Dysplasia, fibrous)
Fibula
  fracture (*see* Fracture, fibula)
  proximal, fracture or dislocation, 250
  baseball, 197
  clubbed, in thyroid acropachy, 311
  mallet, 197
Flatfoot
  flexible, 347
  peroneal spastic (*see* Tarsal coalition)
Flexor digitorum profundus: avulsion of, 199
Fluoroscopy: in carpal instabilities, 214
Fluorosis
  bone abnormalities due to, 317
  differential diagnosis, 314

Follow-up: knowledgeable post-treatment, 2
Fong's disease (*see* Osteo-onychodysostosis)
Foot
  in acromegaly, 311–312
  anatomy, 260–262
  anomalies, congenital, 338–349
    common, 343–349
    key concepts, 338
    terminology, 338–339
  juvenile chronic, 122
  psoriatic, 132
  rheumatoid, 116
  chondroma of, differential diagnosis, 41
  diabetic, osteomyelitis in, 388–389
  forefoot (*see* Forefoot)
  in gout, 163
  hindfoot (*see* Hindfoot)
  in neuropathic joints, 160
  in osteoarthritis, 156
  in Reiter's disease, 136
  rocker-bottom, 346–347
  trauma, 260–264
    key concepts, 260
Forearm, 203–206
  abnormalities, 204
  fracture, 205–206
Forefoot
  normal
    AP view, 344
    lateral view, 345
  pronation, 339
  supination, 339
  valgus, 343
    AP view, 344
    in flatfoot, 347
    lateral view, 345
    in talus, congenital vertical, 348
  varus, 343

  AP view, 344
  in clubfoot, 346
  lateral view, 345
Forestier's disease (*see* Hyperostosis, diffuse idiopathic skeletal)
Fracture
  alignment, 192
  anatomic, 192
  angulation, 192
    valgus, 193
    varus, 193
  ankle, 257–260
    insufficiency, 258
    Maisonneuve, 258
    stress, 258
    triplane, 258–260
  apex, 192
  apposition, 192
    bayonet, 192
    distraction, 192
    lack of, 192
  Barton's, 206
    reverse, 206
  biomechanical principles, 191
  at birth in osteogenesis imperfecta, 331
  Boxer's, 200
  butterfly, 192
  capitate, 211
  carpus, 210–211
  chance, 277–278
  Chauffeur's, 206
  in children, 194–195
  chip, 193
  clavicle, 226
  clay-shoveler's, 275
  closed, 191
  Colles', 205–206
    reverse, 206
  comminution, 192
  complete, 191–192
  complications, 193–194

Fracture (*cont.*)
  condylar, lateral, 223
  definition, 191
  dislocation
    of carpus, 212
    Lisfranc, 160, 263
    pelvic, unstable, 238
    spine, lumbar, 277
  displaced, 192
  elbow
    in adult, 223
    in children, 222-223
  femur
    condyle, 248
    neck, 242
    shaft, 244–245
    stress, 245
  fibula
    proximal, 250
    stress, 250
  forearm, 205–206
  greenstick, 191
  hamate, 211
  hand
    avulsion, 196–200
    shaft, 196
    tuft, 196
  Hangman's, 274
  healing, 193–194
  hip
    epidemiology, 241-242
    fixation, 243
    vertical shear, "open
      book" disruption of,
      240
  humerus, 232–233
    Neer's four-segment
      classification, 232
    proximal, 232
    shaft, 232–233
  Hutchinson's, 206
  imaging modalities in,
    195–196
  incomplete, 191

  intertrochanteric, 242–243
    reduction of, positions
      accepted for, 244
  intracondylar T or Y, 223
  Jefferson's, 273
  Jones', 263
  lunate, 211
  march, 263
  metacarpal, first, 201
    Bennett variety, 201
  metaphyseal
    in scurvy, 312
    transverse, in children,
      205
  metatarsal stress, 263
  oblique, 191
  odontoid, 274
  olecranon, 223
  open, 191
  in osteogenesis imper-
    fecta, 330
  patella, 251
  pathologic, 193
  Pelkin's, in scurvy, 312
  pelvis
    CT of, 235–237
    stable, 238
    unstable fracture-dislo-
      cation, 238
  phalangeal avulsion, sites
    of, 196
  plastic, 191
  position, 192
  pubic, 237
  radius, 205–206
    head, 223
  Rolando, 201
  rotation, 193
  sacrum, 237
  Salter, of radial epi-
    physis, 205
  scaphoid, 210
  scapula, 227
  segmental, 192

Fracture (*cont.*)
  Smith, 206
  spine
    cervical, bilateral verti-
      cal, 273
    cervical, in children,
      276
    thoracic, 276
  spiral, 191
  stress (*see* Stress fracture)
  supracondylar, 222
  teardrop burst, 275
  terminology, 191–196
    accurate terminology to
      be applied in all de-
      scriptions, 191–193
  tibia, 249–250
    plateau, 249
    spine, 250
  Tillaux', juvenile, 259
  transcondylar, 223
  transverse, 191
    in sclerosing dysplasia,
      331
  triquetrum, 211
  trochanter, lesser, avul-
    sion, 244
  volar plate, 199–200
Freiberg's disease, 178-179
Frostbite, 319
  in acro-osteolysis, 395
Fungal septic arthritis, 390

G

Gallium scanning: in tumor
    work-up, 11
Gastrointestinal malabsorp-
    tion: in osteomalacia
    and rickets etiolo-
    gies, 300
Gaucher's disease, 313-314
  avascular necrosis and,
    183
  definition, 313
  epidemiology, 313
  Erlenmeyer flask deform-
    ity in, 313
  key concepts, 313
  osteoporosis in, 313
  radiographic abnormali-
    ties, 313–314
Giant cell tumor, 58–62
  age and, 3
  differential diagnosis, 24
  major, 60
  host response, 60
  key concepts, 58
  in Paget's disease of
    bone, 317
  radiographic work-up, 61
  of tendon sheath, 82
  treatment, 61–62
  WHO classification with
    modification, 15
Giantism, localized, 398–
    399
  definition, 398
  hypervascularity in, 398
  Klippel-Trenaunay-Weber
    syndrome and, 398–
    399
  macrodystrophia lipoma-
    tosa in, 398
  neurofibromatosis in, 398
Gigantism: in neurofibro-
    matosis, 326
Glenohumeral joint
  in osteoarthritis, 152–153
  in spondylitis, 128
Glomus tumor, 74
Gorham's disease, 74
Gout, 161–164
  ankle in, 163
  definition, 161
  differential diagnosis, 164
  epidemiology, 162

Gout (*cont.*)
  extra-articular manifesta-
      tions, 162
  feet in, 163
  hand in, 163
  joints commonly affected,
      163
  key concepts, 161
  knee in, 163
  laboratory tests in, 162
  radiography of, 162–163
  signs, clinical, 162

Hamate
  fracture, 211
  fusion in chondroectoder-
      mal dysplasia, 363
  -triquetrum instability,
      214–216
Hand
  in acromegaly, 311–312
  in arthritis
    juvenile chronic, 122
    psoriatic, 132
    rheumatoid, 112
    rheumatoid, radiogra-
        phy of, 113, 114
  in calcium hydroxyapatite
      deposition disease,
      173
  calcium pyrophosphate
      arthropathy pattern
      in, 167
  chondroma of, differen-
      tial diagnosis, 41
  in CPPD crystal deposi-
      tion disease, 168
  dislocation, 201–203
  fracture (*see* Fracture,
      hand)
  in gout, 163
  in osteoarthritis, 151–152,
      153

wrist in, 163
Granuloma: eosinophilic,
    differential diagno-
    sis, 385
Greenstick fracture, 191
Growth
  disorders, radiation-in-
      duced, 394
  hormone in acromegaly,
      311
  zone disorganization in
      scurvy, 312

H

  trauma, 196–203
    key concepts, 196
Hand-Christian-Schuller
    disease, 99
Hangman's fracture, 274
Hb SS, 381
Herberden's nodes, 150
Hemangioendothelioma,
    75–77
  differential diagnoses,
      major, 77
  key concepts, 76
  metastatic potential, 76
  radiographic work-up, 76
Hemangioma, 71–72
  differential diagnosis,
      major, 72
  key concepts, 71
  radiographic work-up, 72
  treatment, 72, 73
Hemangiopericytoma, 74–
    75
  differential diagnosis,
      major, 75
  radiographic work-up, 75
  treatment, 75
Hemarthroses, 378–379
  in septic arthritis of hip
      in children, 390
Hematologic disorders,
    378–382

Hematopoiesis: in myelofibrosis, 314
Hemochromatosis arthropathy, 169–170
  definition, 169
  differential diagnosis, 170
  epidemiology, 169
  extra-articular manifestations, 169
  joints commonly affected, 170
  key concepts, 169
  laboratory tests, 169
  radiography of, 169-170
  signs, clinical, 169
Hemophilia, 378–380
  A, 378
  arthropathy in, 379
  B, 378
  definition, 378
  epidemiology, 378
  key concepts, 378
  pseudotumor in, 379
  radiographic abnormalities, 378–380
  subperiosteal, in scurvy, 312
Hemosiderin: in hemophilia, 379
Heparin: causing bone abnormalities, 317
Heterotopic bone: in septic arthritis, 391
Hilgenreiner's line, 333, 334
Hindfoot
  calcaneus, 339, 340
  equinus, 339, 340
    in clubfoot, 346
    in talus, congenital vertical, 348
  normal, 339, 340
    AP view, 342
    lateral view, 341
  valgus, 339–343

    AP view, 342
    in flatfoot, 347
    in talus, congenital vertical, 348
  varus, 339–343
    AP view, 342
    in clubfoot, 346
Hip
  anatomy, 239–241
  in arthritis
    juvenile chronic, 123
    rheumatoid, 116
    septic, in children, 390
    septic, in children, differential diagnosis, 390
  arthroplasty (*see* Arthroplasty, total hip)
  in CPPD crystal deposition disease, 168
  dislocation (*see* Dislocation, hip)
  fracture (*see* Fracture, hip)
  groin, lateral view of, 241
  normal, 334
  normal AP of, 247
  in osteoarthritis, 154–155
  in osteochondrosis, 178
  in spondylitis, 128
  subluxation, lateral, 333
  trauma, 239–247
    key concepts, 239
Histamine release: in mastocytosis, 313
Histiocytoma
  differential diagnosis, 65
  malignant fibrous, 82–84
    age and, 3
    differential diagnosis, major, 84
    key concepts, 83
    radiographic work-up, 84
    treatment, 84

Histiocytosis: differential
diagnosis, 64–65
Histiocytosis X, 96–99
differential diagnosis,
major, 98
key concepts, 97
radiographic work-up, 98
treatment, 98–99
Hodgkin's disease, 67–68
differential diagnosis,
major, 68
key concepts, 67
Homocystinuria, 295, 328
collagen synthesis defect
in, 328
contracture in, flexion,
328
cystathionine synthetase
in, 328
definition, 328
epidemiology, 328
key concepts, 328
musculoskeletal features,
328
nonmusculoskeletal fea-
tures, 328
osteoporosis in, 328
pectus excavatum in, 328
vertebral body scalloping
in, 328
Hormones: GH in acrome-
galy, 311
Host response
adamantinoma, 89
in Ewing's sarcoma, 64
giant cell tumor, 60
osteosarcoma, 29
tumors and, 5
Humerus
anterior humeral lines on
true lateral film, 220
in chondrodysplasia calcificans punctata, 365
fracture (*see* Fracture,
humerus)

separation, in children,
224
Hutchinson's fracture, 206
Hydroxyproline: in Paget's
disease of bone, 315
Hypercorticism: differential
diagnosis, 295–296
Hyperelastic skin: in Eh-
lers-Danlos syn-
drome, 329
Hyperemia: in hemophilia,
379
Hyperirritability: and periosteal reaction in in-
fant, 396
Hyperostosis
cortical, 396
diffuse idiopathic skele-
tal, 185–187
definition, 185
differential diagnosis,
187
epidemiology, 185
key concepts, 185
radiography of, 186
signs, clinical, 185–186
spine in, 186–187
from meningioma, differ-
ential diagnosis, 18
Hyperparathyroidism, 302–
306
Brown tumors in, 306
calcifications in, 306
chondrocalcinosis in, 306
clinical findings, 303
etiology, 303
key concepts, 303
laboratory values, 303
osteopenia due to, 293
radiographic findings,
303–306
resorption in, 305
sclerosis in, 305–306

Hyperparathyroidism
(*cont.*)
  secondary, in renal osteo-
    dystrophy, 307
  skeletal manifestations,
    303–306
Hyperplasia: marrow, in
    anemia, 380, 381–382
Hyperthyroidism, 295, 310–
    311
  in children, 311
  myopathy in, 311
  osteoporosis in, 311
Hypertrophic
  neuropathic joints, 159
  nonunion, 194
  osteoarthropathy (*see* Os-
    teoarthropathy,
    hypertrophic)
Hypervascularity: in local-
    ized giantism, 398
Hypervitaminosis A: in in-
    fant, 397
Hypocalcemia: in pseudo-
    hypoparathyroidism,
    309
Hypochondroplasia, 358–
    359
  clinical features, 358
  diagnosis, differential,
    358–359
  inheritance, 358

Ilia: flared, in asphyxiating
    thoracic dysplasia,
    362
Iliac neck and wing: in mu-
    copolysaccharidoses,
    376
Iliac wings
  in achondroplasia, 357
  central chondrosarcoma
    in, 53

prognosis, 358
progression, 358
radiographic features, 358
synonyms, 358
Hypoparathyroidism, 309–
    310
  calcification in, 309
  definition, 309
  features distinguishing
    from pseudohypo-
    parathyroidism, 310
  hypoplastic dentition in,
    309
  key concepts, 309
  osteosclerosis, 309
  radiographic abnormali-
    ties, 309
Hypophosphatasia: mimick-
    ing rickets, 302
Hypoplasia (*see* Odontoid,
    hypoplastic)
Hypoplastic dentition: in
    hypoparathyroidism,
    309
Hypothyroidism, juvenile,
    310
  epiphyses in, 310
  femoral capital epiphysis
    in, 310
  stippled epiphysis in, 310
  wormian bones in, 310
Hypovitaminosis C (*see*
    Scurvy)

I

Ilium: in spondyloepiphy-
    seal dysplasia con-
    genita, 367
Imaging
  bone (*see* Bone, imaging)
  of fractures, 195–196
  magnetic resonance (*see*
    Magnetic resonance
    imaging)

Imaging (*cont.*)
  in osteomyelitis, 387
  in tumor work-up, 11
Implant: carpal, in total
      knee arthroplasty,
      291
Incomplete fracture, 191
Indium-111 imaging: in os-
      teomyelitis, 387
Infant
  coxa vara (*see* Coxa vara,
      infantile)
  hypervitaminosis A, 397
  osteomyelitis, hematoge-
      nous spread, 383–
      384
  periosteal reaction (*see*
      Periosteal reaction,
      in infant)
  rickets, 397
  scurvy, 397
Infarction
  in anemia, 380, 381

diaphyseal, 295
Infection, 382–393
  key concepts, 382
  organisms, 382
  soft tissue calcification
      and, 400
  spine (*see* Spine,
      infection)
Intertrochanteric fracture,
      242–243
  reduction of, positions
      accepted for, 244
Intoxication: aluminum, in
      renal osteodystro-
      phy, 307
Intracondylar T or Y frac-
      ture, 223
Intracortical resorption: in
      hyperparathy-
      roidism, 305
Involucrum: in osteomyeli-
      tis, 385
Iridocyclitis, 120

J

Jaccoud's arthropathy, 148
Jaw
  osteosarcoma of, 38
  in slcerosing dysplasia,
      332
Jefferson's fracture, 273
Joint
  AC, 226–227
  in arthritis, septic, 389
  carpometacarpal, articula-
      tion, 202
  Charcot (*see* Neuropathic
      joints)
  Charcot-like, 295
  contracture, in dias-
      trophic dwarfism,
      364
  enlargement in hemophi-
      lia, 379

glenohumeral (*see* Gleno-
      humeral joint)
  in hemophilia, 379
  of Luschka, 157
  neuropathic (*see* Neuro-
      pathic joints)
  in osteomyelitis, hema-
      togenous spread, 384
  sacroiliac (*see* Sacroiliac
      joint)
  shoulder, anatomy, 227–
      228
  sternoclavicular, 225–226
  subtalar, Harris-Beath
      view, 352
  talocalcaneal, 262
  talonavicular, dislocation,
      348
  tarsometatarsal, normal
      alignment, 261

Jones' fracture, 263

Keinbock's disease, 178
Kidney
  necrosis, papillary, in
    anemia, 381
  osteodystrophy (*see* Os-
    teodystrophy, renal)
  tubular disorders, osteo-
    malacia and rickets,
    301
Klippel-Trenaunay-Weber
  syndrome: and local-
  ized giantism, 398–
  399
Knee
  anatomy, 254–256
  in arthritis, juvenile
    chronic, 122–123
  arthrography, 253–254
  arthroplasty (*see* Arthro-
    plasty, total knee)
  arthroscopy of, 253
  in CPPD crystal deposi-
    tion disease, 166–167
  dislocation, 252–253

Lead: bone abnormalities
  due to, 317
Legg-Calvé-Perthes disease,
  178, 390
Lens dislocation: in homocystinuria, 328
Letterer-Siwe disease, 99
Ligaments
  collateral, 200
  in Ehlers-Danlos syn-
    drome, 329
  radial, 200
  ulnar, 200
Ligamentum teres: elon-
  gated in hip disloca-
  tion, congenital, 335

Judet, 235
Judet view: of pelvis, 236

K

  epiphyseal injury, 250–
    251
  in gout, 163
  in hemarthroses, 379
  in hemophilia, 379
  MRI of, 253
  in neuropathic joints, 160
  in osteoarthritis, 155
  in osteochondrosis, 179
  osteosarcoma of, 28, 33
  rheumatoid arthritis of,
    116
    radiography of, 117
  trauma, 247–254
    key concepts, 248
Kohler's disease, 180
Kyphoscoliosis
  in Ehlers-Danlos syn-
    drome, 329
  in Marfan's syndrome,
    327
  in neurofibromatosis, 325
Kyphosis: angular, in tu-
  berculosis of spine,
  392

L

Limb (*see* Extremities)
Limbus "thorn": inverted,
  in hip dislocation,
  congenital, 335
Lipoblastomatosis, 81
Lipoma, 81
  osseous, 82
  soft tissue, 81
Lipomatosis, 81
Liposarcoma, 85
Lisfranc fracture dislocation, 160, 263
Liver: abnormalities in osteomalacia and rick-
  ets etiologies, 300–
  301

Lordosis: cervical, 264
Lunate
-capitate instability, 214
dislocation, 213–214
fracture, 211
malacia, 178
normal PA view, 207
volar flexion of, 216
Lupus erythematosus, systemic, 139–141
definition, 139
differential diagnosis, 141
epidemiology, 139
extra-articular manifestations, 139–140
joints commonly affected, 140

key concepts, 139
laboratory tests, 139
radiography of, 140
signs, clinical, 139
symmetry in, 140–141
Luschka: joints of, 157
Lymphangioma, 72–73
differential diagnosis, major, 72–73
radiographic work-up, 73
Lymphoma
age and, 3
of bone, 66–67
differential diagnosis, major, 66
radiographic work-up, 67
treatment, 67

M

Macrocephaly, 374
Macrodystrophia lipomatosa, 81
in localized giantism, 398
Maffucci's syndrome, 43–44
Magnetic resonance imaging
of knee, 253
in tumor work-up, 12
Malabsorption: gastrointestinal, in osteomalacia and rickets etiologies, 300
Mallet finger, 197
Mandible
in anemia, 380
in periosteal reaction in infant, 396
March fracture, 263
Marfan's syndrome, 326–327
arachnodactyly in, 327
definition, 327
epidemiology, 327
key concepts, 327

kyphoscoliosis in, 327
musculoskeletal features, 327
nonmusculoskeletal features, 327
pectus excavatum in, 327
spondylolysis in, 327
vertebral body scalloping in, 327
Marrow
host reaction in mastocytosis, 313
hyperplasia in anemia, 380, 381–382
sclerotic, in myelofibrosis, 314
tumors, 62–71
WHO classification with modification, 15
Mast cell disorder: in mastocytosis, 312
Mastocytosis, 312–313
definition, 312–313
differential diagnosis, 314
histamine release in, 313

Mastocytosis (*cont.*)
  host reaction to marrow
    in, 313
  radiographic abnormalities, 313
  sclerosis in, bony, 313
Meningioma: hyperostosis
    from, differential diagnosis, 18
Meniscus
  chondrocalcinosis, 166
  discoid, 253
Mental retardation
  in homocystinuria, 328
  in Morquio's syndrome, 376
Mesenchymoma: malignant, 86
Mesomelia, 371
Metabolic bone disease, 293–319
  laboratory values, serum, 319
Metacarpal
  in chondrodysplasia calcificans punctata, 365
  dislocation, 201–203
  first
    in diastrophic dwarfism, 364
    fracture, 201
    fracture, Bennett variety, 201
  in mucopolysaccharidoses, 376
  short, in pseudohypoparathyroidism, 310
  in thyroid acropachy, 311
Metaphyseal chondrodysplasia: Schmid type, mimicking rickets, 302
Metaphysis
  in asphyxiating thoracic dysplasia, 362

  in chondrodysplasia calcificans punctata, 365
  in dwarfism, thanatophoric, 360
  flared, in achondroplasia, 357
  fracture
    in scurvy, 312
    transverse, in children, 205
  line: in scurvy, 312
  in osteomyelitis, 384
    in children, 385
    syphilis, 388
  tumors in, 4
Metastases, 16
  blastic, differential diagnosis, 19
  bone, 102–105
    appearance of, 103–104
    differential diagnosis, 104–105
    frequency, 102
    imaging, 102–103
    location, 104
    radiography of, 103
  diagnosis, differential, 314
  of myeloma, and age, 3
  of neuroblastoma, and age, 3
  vascular
    angiography of, 103
    embolization, of, 103
Metatarsals, 260
  in pseudohypoparathyroidism, 310
  stress fracture, 263
Midcarpal dislocation, 212–213
Milwaukee shoulder, 173
Monostotic tumors, 5

Morquio's syndrome, 376–377
Mucopolysaccharidoses, 373–377
  radiographic findings, 374–377
Myelofibrosis, 314
  definition, 314
  diagnosis, differential, 314
  radiographic findings, 314
Myeloma, multiple, 68–71
  age and, 68
  differential diagnosis, major, 69–70
  key concepts, 68
  location, 69

metastases of, and age, 3
osteopenia due to, 294
pattern, 68
radiographic work-up, 70
size, 69
treatment, 70–71
Myopathy: in hyperthyroidism, 311
Myositis ossificans, 101–102, 281–285
  with burns, 283
  differential diagnosis, 29–30, 34, 45, 283
  in elbow dislocation, 224
  juxtacortical, 281–282
  key concepts, 281
  with neurologic disorders, 283
  progressiva, 284–285

N

Nail-patella syndrome (*see* Osteoonychodysostosis)
Navicular trauma, 263
Necrosis
  aseptic, 295
    in renal osteodystrophy, 308
  avascular, 182–184
    definition, 182–183
    differential diagnosis, 184
    in hip dislocation, congenital, 335
    key concepts, 182
    radiography of, 183–184
    sites of common occurrence, 184
  renal papillary, in anemia, 381
  vertebral body, in anemia, 381
Neer's four-segment classification: of humerus fracture, 232

Neoplasm (*see* Tumors)
Neuroarthropathy
  shoulder, 160
  spine, 160–161
Neuroblastoma: metastatic, and age, 3
Neurofibroma: in neurofibromatosis, 326
Neurofibromatosis, 325–326
  exostoses in, 326
  in giantism, localized, 398
  gigantism in, 326
  key concepts, 325
  kyphoscoliosis in, 325
  rib in, 326
  scoliosis and, 322
  skeletal involvement, 325
  skull in, 326
  stripped periosteum in, 325–326
  tibia in, 326

Neurologic disorders: with myositis ossificans, 283
Neuropathic joints, 158–161
  atrophic, 159
  definition, 158–159
  differential diagnosis, 161
  extra-articular manifestations, 159
  feet in, 160
  five Ds in, 159
  hypertrophic, 159
  joints commonly affected, 160
  key concepts, 158
  knee in, 160
  radiography of, 159–160
  signs, clinical, 159
Newborn: rickets, 301–302
Niemann-Pick disease, 314
Nonunion, 194
  atrophic, 194
  hypertrophic, 194
Nursemaid's elbow, 224

O

Oblique fracture, 191
Ochronosis (*see* Alkaptonuria)
O'Donoghue's terrible triad, 253
Odontoid
  erosion in rheumatoid arthritis, 118
  fracture, 274
  hypoplastic
    in Morquio's syndrome, 376
    in spondyloepiphyseal dysplasia congenita, 367
  process, 267
Older age (*see* Osteosarcoma, in aged)
Olecranon fracture, 223
Ollier's disease, 43
Open fracture, 191
Orthopedic hardware, 285–291
  key concepts, 285
Os
  odontoideum, 272–273
  terminale, 271–272
    variants, spectrum of, 272
Osgood-Schlatter's disease, 179
Ossification: spine, in achondrogenesis, 361
Osteoarthritis, 149–158
  definition, 149
  differential diagnosis, 158
  in Ehlers-Danlos syndrome, 329
  epidemiology, 149–150
  feet in, 156
  glenohumeral joint in, 152–153
  hand in, 151–152, 153
  hip in, 154–155
  in hip dislocation, congenital, 335
  joints commonly affected, 151–157
  key concepts, 149
  knee in, 155
  in Paget's disease of bone, 317
  radiography of, 150–151
    survey films, 158
  signs, clinical, 150
  spine in, 156–157
  symmetry in, 157
  true, 157
  wrist in, 152, 153
Osteoarthropathy, hypertrophic, 181–182
  definition, 181

Osteoarthropathy, hyper-
trophic (*cont.*)
diagnosis, differential,
182
key concepts, 181
Osteoblast: in Paget's disease of bone, 315
Osteoblastoma, 22–25
differential diagnosis,
major, 24
key concepts, 23
radiography of, 24
in spine, 23
treatment, 25
Osteochondritis dissecans,
179
talar dome, 180
Osteochondroma, 44–47
(*See also* Exostoses)
differential diagnosis, 34,
283
major, 45
key concepts, 44–45
multiple hereditary exos-
toses, 46
radiographic work-up, 46
treatment, 46
Osteochondrosis, 177–181
definition, 177
differential diagnosis, 181
epidemiology, 177
hip in, 178
joints affected, 178
key concepts, 177
knee in, 179
radiography of, 177–178
tibial apophysis in, 179
Osteoclast
in Paget's disease of
bone, 315
in sclerosing dysplasia,
331
Osteodystrophy, renal,
306–309

abnormalities following
dialysis, 308–309
aluminum intoxication in,
307
arthritis and, septic, 308
calcifications in, 308
carpal tunnel syndrome
in, 308
features, 307
hyperparathyroidism in,
secondary, 307
key concepts, 306
laboratory values, 307
necrosis in, aseptic, 308
osteomalacia in, 301, 307
osteomyelitis and, 308
osteosclerosis in, 307
radiographic features,
307–309
resorption in, 307
rickets in, 301, 307
slipped epiphyses in, 307
spondyloarthropathy in,
308–309
Osteogenesis imperfecta,
296, 329–331
congenita, 330–331
definition, 330
epidemiology, 330
features, 330
key concepts, 330
tarda 1, 331
tarda 2, 331
Osteoid (*see* Osteoma,
osteoid)
Osteolysis
acro-osteolysis (*see* Acro-
osteolysis)
massive, 74
Osteoma, 17–18
differential diagnosis,
major, 18
key concepts, 17
Osteoma, osteoid, 19–22
clinical features, 21

Osteoma, osteoid (*cont.*)
    differential diagnosis, 19,
        24
        major, 21–22
    in femoral neck, 20
    host response in, 21
    key concepts, 19
    in osteomyelitis, 386
    radiography in, 22
    in scoliosis, 322
    treatment, 22
Osteomalacia, 298–302
    definition, 298–299
    etiologies, 300–302
    key concepts, 298
    laboratory findings, algo-
        rithm of, 298
    metabolism in, 299–300
    in osteodystrophy, renal,
        307
    osteopenia due to, 293
    radiographic signs, 299
Osteomyelitis
    in anemia, 381
    anemia confused with,
        380
    chronic, 386
        host reaction in, 386
    complications, 387
    in diabetic foot, 388–389
    diagnostic difficulties,
        386-387
    differential diagnosis,
        64,65
    hematogenous spread,
        383–384
        in adult, 384
        in children, 384
    imaging, 387
    pin tract, 389
    radiographic appearance,
        384–386
    renal osteodystrophy
        and, 308

    routes of involvement,
        382–383
        in infant, 383–384
    special cases of, 387–389
    subacute, 385–386
        diagnosis, differential,
            385
    syphilis, 388
Osteonecrosis
    radiation, 394
    spontaneous, 249
Osteo-onychodysostosis,
        354
    definition, 354
    radiographic abnormali-
        ties, 354
Osteopenia
    in anemia, 380
    causes, 293
    in hyperparathyroidism,
        303–305
    in scurvy, 312
Osteopetrosis: in sclerosing
        dysplasia, 331
Osteophytes: in osteoarthri-
        tis, 150
Osteopoikilosis: in scleros-
        ing dysplasia, 332
Osteoporosis, 293–297
    in arthritis
        juvenile chronic, 121
        rheumatoid, 111
    bone mass quantitative
        analysis, 297
    circumscripta in Paget's
        disease of bone, 316
    CT of, 297
    definition, 293–294
    diagnosis, differential,
        295–296
    drug-related, 296
    dual-photon absorptiom-
        etry in, 297

Osteoporosis (*cont.*)
etiologies, 294–296
in Gaucher's disease, 313
generalized, 294–296
in hemophilia, 379
in homocystinuria, 328
in hyperthyroidism, 311
idiopathic juvenile, 296
involutional, 294–295
key concepts, 293
localized, 296–297
in osteogenesis imper-
fecta, 330
in Paget's disease of
bone, 316
radiographic findings,
294–296
senile, 294–295
transient of hip, 297
transient regional, 187,
296–297
migratory, 297
Osteosarcoma, 27–41
age and, 3
in aged, 38–39
location of, 38
radiographic work-up,
39
survival, 39
treatment, 39
conventional, 27–31
radiographic work-up,
30
treatment, 30–31
differential diagnosis, 24,
35–36, 283
major, 29–30
extraskeletal, 86
host response, 29
intraosseous, low-grade,
36–38

Paget's disease of bone,
315–317

differential diagnosis,
36–38
radiographic work-up,
38
treatment, 38
of jaw, 38
key concepts, 27
of knee, 28
multicentric, 39–40
parosteal, 32–35
differential diagnosis,
34, 45, 283
of knee, 33
radiographic work-up,
34
treatment, 35
periosteal, 35–36
differential diagnosis,
major, 35, 283
radiographic work-up,
36
treatment, 36
after radiotherapy, 39
soft tissue, 40
telangiectatic, 31–32
differential diagnosis,
major, 32
radiographic work-up,
32
Osteosarcomatosis, 39–40
Osteosclerosis
in hypoparathyroidism,
309
in pseudohypoparathy-
roidism, 309
in renal osteodystrophy,
307
Osteotomy: in hip dislocation, congenital, 335–
336

P

clinical signs, 315
complications, musculo-
skeletal, 316–317

Paget's disease of bone
(*cont.*)
cranium in, 316
definition, 315
diagnosis, differential,
314
epidemiology, 315
fracture in, stress, 316
giant cell tumors in, 317
key concepts, 315
laboratory abnormalities,
315
lytic, 315–316
mixed, 316
osteoarthritis in, 317
pelvis in, 316
radiographic appearance,
315–316
sclerotic, 316
sites of involvement, 316
spine in, 316
tubular bones in, 316
Panner's disease, 180
Parham bands: in total
knee arthroplasty,
290
Parosteal (*see* Osteosarcoma, parosteal)
Patella
complicating total knee
arthroplasty, 289
dislocation, 251–252
fracture, 251
in gout, 163
in hemophilia, 379
nail-patella syndrome (*see*
Osteo-
onychodysostosis)
tilt and displacement,
sunrise view of, 252
trauma, 251
Pauciarticular disease, 120
Pectus excavatum
in homocystinuria, 328

in Marfan's syndrome,
327
Pelkin's fracture: in scurvy,
312
Pelvis
anatomy, 233-237
AP, 234
fracture (*see* Fracture,
pelvis)
in hemophilia, 379
Judet view of, 236
in Paget's disease of
bone, 316
trauma, 233-239
key concepts, 233
Pemberton acetabuloplasty:
in hip dislocation,
congenital, 335
Perilunate dislocation, 212
Periosteal desmoid, 80–81
Periosteal osteosarcoma (*see*
Osteosarcoma,
periosteal)
Periosteal reaction
in anemia, 380
in hemophilia, 380
in infant, 396–398
key concepts, 396
onset anytime during
infancy, 397–398
onset after 6 months,
397
onset before 6 months,
396
physiologic periosteal
new bone, 396
prostaglandin and, 398
"wavy" radius differ-
entiated from, 398
in osteomyelitis, 384
syphilis, 388
in septic arthritis, 391
to tumor, 397–398

Periosteum: stripped in neurofibromatosis, 325–326
Periostitis
in juvenile chronic arthritis, 122
posttraumatic, 282–283
in spondylitis, 126
Peroneal spastic flatfoot (*see* Tarsal coalition)
Pes
cavus, 347–349
etiology, 347–349
planovalgus, 347
Phagocytes: in osteomyelitis, 384
Phalanx
avulsion
fracture, sites of, 196
of volar fragment, 199–200
dislocation, 201
middle, AP view of extensor mechanism, 198
in pseudohypoparathyroidism, 310
terminal, glomus tumor of, 74
in thyroid acropachy, 311
Phenobarbital: osteomalacia and rickets, 301
Phosphate, 318
Photon: dual-photon absorptiometry in osteoporosis, 297
Pigmented synovitis (*see* Synovitis, pigmented)
Pin tract osteomyelitis, 389
Planus, 338
Plasmacytoma, 69
radiotherapy of, 71
Plastic fracture, 191

Platyspondyly
in Morquio's syndrome, 376
in thanatophoric dwarfism, 360
Pneumatized sinuses: in acromegaly, 311
Poisoning
vitamin A, causing bone abnormalities, 317
vitamin D, causing bone abnormalities, 317
Polyarthritis: of unknown etiology, 110–138
Polyarticular disease
seronegative, 121
seropositive, 121
Polydactyly: in chondroectodermal dysplasia, 363
Polymyositis/dermatomyositis, 144–145
definition, 144
differential diagnosis, 145
epidemiology, 144
extra-articular manifestations, 144
joints commonly affected, 145
key concepts, 144
laboratory tests, 144
radiography of, 144–145
signs, clinical, 144
Polyostotic tumors, 5
Polyvinylchloride, 319, 395
Pott's disease, 392
Prematurity: copper deficiency in, 302
Prostaglandins
bone abnormalities due to, 317
periosteal reaction in infant and, 398

Prosthesis: endoprosthesis
in total hip arthro-
plasty, 288
Pseudarthrosis, 338
definition, 338
Pseudoachondroplasia, 359
clinical features, 359
diagnosis, differential,
359
inheritance, 359
prognosis, 359
progression, 359
radiographic features, 359
synonyms, 359
Pseudogout, 165
Pseudohypopara-
thyroidism, 309–310
calcification in, 309–310
etiology, 309
features distinguishing
from hypoparathy-
roidism, 310
hypocalcemia in, 309

Radiation
avascular necrosis after,
183
-induced abnormalities,
394–395
osteonecrosis, 394
Radiocapitellar line, 221
Radiography
achondrogenesis, 361
achondroplasia, 357
acromegaly, 311–312
adamantinoma, 89–90
in alkaptonuria, 174
in amyloidosis, 146–147
anemia, 380
angiomatosis, 74
angiosarcoma, 77
of arthritis

osteosclerosis in, 309
Pseudomonas: in drug abuse,
382
Pseudopseudohypopara-
thyroidism, 310
radiographic abnormali-
ties, 310
Pseudotumor: in hemophi-
lia, 379–380
Psoas abscess: calcified, in
tuberculosis of spine,
392
Psoriatic
arthritis (*see* Arthritis,
psoriatic)
arthropathy, 121
Pubic fracture, 237
Pulmonary infarction: in
anemia, 381
Pulvinar: in hip dislocation,
congenital, 335
Pyknodysostosis: in sclerosing dysplasia, 332
Pyle's disease, 314

R

juvenile chronic, 121–
122
juvenile chronic, sur-
vey films for early
diagnosis, 124
psoriatic, 131–132
psoriatic, survey films
for early diagnosis,
134–135
septic, 389
septic, sacroiliac joint,
391
rheumatoid, 111–112
rheumatoid, survey
films for early diag-
nosis, 119
arthrogryposis multiplex
congenita, 324–325
in battered child syn-
drome, 279–280

Radiography (*cont.*)
of bone metastases, 103
in calcium hydroxyapatite deposition disease, 172
chest, in tumor work-up, 13
chondroblastoma, 49
chondroma, 42
juxtacortical, 52
in chondromatosis, synovial, 176–177
chondrosarcoma
central, 55
clear cell, 57
mesenchymal, 58
peripheral, 56
chordoma, 88
cleidocranial dysostosis, 354
clubfoot, 345–346
in CPPD crystal deposition disease, 166
survey films, 168
cyst, bone
aneurysmal, 95
solitary, 92
description of tumor on radiograph, 2
discitis, 392
dislocation hip, congenital, 333
dwarfism
classification, approach to, 368–373
diastrophic, 364
thanatophoric, 360
dyschondrosteosis, 366–367
dysplasia
asphyxiating thoracic, 362
fibrous, 101
in enostosis, 19

of femoral epiphysis, slipped capital, 246
femoral focal deficiency, proximal, 336–337
fibroma
chondromyxoid, 50
desmoplastic, 80
nonossifying, 96
ossifying, 26
Gaucher's disease, 313–314
giant cell tumor, 61
in gout, 162–163
hemangioendothelioma, 76
hemangioma, 72
hemangiopericytoma, 75
in hemochromatosis arthropathy, 169–171
hemophilia, 378–380
histiocytoma, malignant fibrous, 84
histiocytosis X, 98
in hyperostosis, diffuse idiopathic skeletal, 186
hyperparathyroidism, 303–306
hypochondroplasia, 358
hypoparathyroidism, 309
in lupus erythematosus, systemic, 140
lymphangioma, 73
lymphoma of bone, 67
mastocytosis, 313
metatarsus adductus, 349
mucopolysaccharidoses, 374–377
myelofibrosis, 314
myeloma, multiple, 70
in necrosis, avascular, 183–184
in neuropathic joints, 159–160

Radiography (*cont.*)
  in osteoarthritis, 150–151
    survey films, 158
  of osteoblastoma, 24
  osteochondroma, 46
  in osteochondrosis, 177–178
  osteodystrophy, renal, 307–309
  in osteoma, osteoid, 22
  osteomalacia, 299
  osteomyelitis, 384–386
  osteo-onychodysostosis, 354
  osteoporosis, 294–296
  osteosarcoma
    in aged, 39
    conventional, 30
    intraosseous, low-grade, 36
    parosteal, 34
    periosteal, 36
    telangiectatic, 32
  Paget's disease of bone, 315–316
  in polymyositis/dermatomyositis, 144–145
  pseudoachondroplasia, 359
  pseudopseudohypoparathyroidism, 310
  in Reiter's disease, 135–136
    survey films for early diagnosis, 138
  of rheumatic fever, 148
    survey films, 149
  rickets, 299
  sarcoma
    Ewing's, 65
    synovial cell, 86
  in scleroderma, 142–143
  scoliosis, 323
  spinal infection, 392

  spinal instability, cervical, 275
  of spondylitis, 125–126
    survey films for, 130
  of synovitis, pigmented villonodular, 175–176
  talus, congenital vertical, 347
  tarsal coalition, 350
  in tumor work-up, 11
  in Wilson's disease, 171
  of wrist (*see under* Wrist)
Radionuclide studies (*see* Imaging)
Radiotherapy
  osteosarcoma after, 39
  of plasmacytoma, 71
  scoliosis and, 323
Radioulnar dislocation, 205
Radius
  deviation, 208
  dislocation of head, 224
  distal, normal ulnar tilt of articular surface on PA film, 204
  in dyschondrosteosis, 367
  epiphysis, Salter fracture of, 205
  fracture, 205–206, 223
  in hemophilia, 379
  "wavy," differentiated from periosteal reaction in infant, 398
Reflex sympathetic dystrophy, 296
Reiter's disease, 135–138
  definition, 135
  differential diagnosis, 138
  epidemiology, 135
  extra-articular manifestations, 135
  feet in, 136
  joints most commonly affected, 136–138
  key concepts, 135

Reiter's disease (*cont.*)
  laboratory tests, 135
  radiography of, 135–136
    survey films for early
      diagnosis, 138
  signs, clinical, 135
  spine in, 138
  symmetry in, 138
Resection: surgical, guidance of, 2
Reticulum cell sarcoma (*see*
  Lymphoma, bone)
Rhabdomyosarcoma, 86
Rheumatic fever, 147–149
  definition, 147
  differential diagnosis,
    148–149
  joints commonly affected,
    148
  key concepts, 147
  laboratory tests, 148
  radiography of, 148
    survey films, 149
  signs, clinical, 147–148
  symmetry, 148
Rheumatoid arthritis, 110–
    124
  ankylosis in, 112
  cartilage destruction in,
    111
  cysts in, subchondral,
    112
  definition, 110
  differential diagnosis, 119
  dot-dash pattern in, 111
  of elbow, 113
  epidemiology, 110
  erosive changes, 111
    subchondral erosions,
      111
  extra-articular manifesta-
    tions, 111
  of feet, 116
  of hand, 112

  radiography of, 113,
    114
  of hip, 116
  joint distribution, com-
    mon, 112
  juvenile chronic, 120–124
    definition, 120–121
    key concepts, 120
    radiography of, 121–
      122
  key concepts, 110
  of knee, 116
    radiography of, 117
  laboratory tests, 110–111
  osteoporosis in, 111
  radiography of, 111–112
    survey films for early
      diagnosis, 119
  rotator cuff tear in, 114–
    115
  of sacroiliac joint, 116
  of shoulder, 114–115
    radiography of, 115
  signs, clinical, 110
  of spine, 116–119
  symmetry in, 119
  of wrist, 112
Rhizomelia, 370
Rhizomelic tubular bone: in
    thanatophoric dwarf-
    ism, 360
Rhomboid fossa, 226
Rib
  in achondrogenesis, 361
  in asphyxiating thoracic
    dysplasia, 362
  in chondroectodermal
    dysplasia, 363
  in dwarfism, thanato-
    phoric, 360
  in mucopolysacchari-
    doses, 374
  in neurofibromatosis, 326

Rickets, 298–302
definition, 298–299
etiologies, 300–302
in infant, 397
key concepts, 298
laboratory findings, algorithm of, 298
metabolism in, 299–300
in newborn, 301–302
in osteodystrophy, renal, 307
radiographic signs, 299
tumor-associated, 301–302

vitamin D resistant, 301–302
Ring sequestrum: in pin tract osteomyelitis, 389
Rocker-bottom foot, 346–347
Rod: intermedullary, in total knee arthroplasty, 290
Rolando fracture, 201
Rotator cuff tear, 230–231
in rheumatoid arthritis, 114–115

S

Saber shin deformity, 388
Sacroiliac joint
rheumatoid arthritis of, 116
in septic arthritis, 391
Sacrum, 234
fracture, 237
Salmonella
in anemia, 381
sickle cell and, 382
Salter fracture: of radial epiphysis, 205
Salter opening wedge osteotomy: in hip dislocation, congenital, 336
Sarcoidosis, 393–394
definition, 393
diagnosis, differential, 394
epidemiology, 393–394
key concepts, 393
osseous manifestations, 394
Sarcoma
angiosarcoma (*see* Angiosarcoma)
chondrosarcoma (*see* Chondrosarcoma)

Ewing's, 62–65
age and, 3
differential diagnosis, 29
differential diagnosis, major, 64–65
host response, 64
key concepts, 62
radiographic work-up, 65
treatment, 65
fibrosarcoma, 82–84
liposarcoma, 85
osteosarcoma (*see* Osteosarcoma)
reticulum cell (*see* Lymphoma, of bone)
soft tissue, surgical staging, 6
synovial cell, 85–86
calcification in, 86
radiography of, 86
Scanning (*see* Imaging)
Scaphoid
fracture, 210
normal PA view, 207
Scapholunate dissociation, 214, 216

Scapula, 227
  fracture, 227
Scheuermann's disease, 180
Schmorl's nodes, 156
Schwannoma: malignant,
  86
Sclerae: blue, in osteogenesis imperfecta, 330
Scleroderma, 141–144
  definition, 141
  differential diagnosis,
    143–144
  epidemiology, 141
  extra-articular manifesta-
    tions, 142
  joints commonly affected,
    143
  key concepts, 141
  laboratory tests, 142
  radiography in, 142–143
  signs, clinical, 141–142
  symmetry in, 143
Sclerosing (*see* Dysplasia,
  sclerosing)
Sclerosis
  in alkaptonuria, 174
  bony, in mastocytosis,
    313
  in discitis, 393
  discogenic, 156, 393
  of femoral component in
    total hip arthro-
    plasty, 287
  in hyperparathyroidism,
    305–306
  in necrosis, avascular,
    183–184
  in osteomyelitis, 386
  progressive systemic (*see*
    Scleroderma)
  segmental, idiopathic,
    156
  in septic arthritis, 389
  subchondral, in osteoar-
    thritis, 150–151

Sclerotic
  epiphyseal rim: in
    scurvy, 312
  marrow in myelofibrosis,
    314
  Paget's disease of bone,
    316
Scoliosis, 320–324
  congenital, 322–323
  definition, 321
  etiology, 321–323
  in homocystinuria, 328
  idiopathic, 321
  key concepts, 321
  radiography of, 323
  radiotherapy and, 323
  treatment, 323–324
  tumors and, 322
Scorbutic rosary, 312
Screws: dynamic, in total
  knee arthroplasty,
  290
Scurvy, 312
  calcification in, 312
  collagen in, 312
  definition, 312
  hemorrhage in, subper-
    iosteal, 312
  in infant, 397
  metaphyseal corner frac-
    tures in, 312
  metaphyseal line in, 312
  osteopenia in, 312
  radiographic abnormali-
    ties, 312
  sclerotic epiphyseal rim
    in, 312
Sella: enlargement in acromegaly, 311
Septic (*see* Arthritis, septic)
Sequestrum
  in osteomyelitis, 385
  ring, in pin tract osteo-
    myelitis, 389

Serratia: in drug abuse, 382
Sesamoid bones, 262
Shenton's line, 333
Short stature,
    disproportionate
    short-limbed, normal
        trunk, 370–372
    short limbs and short
        trunk, 374
    short trunk and spinal in-
        volvement, 373
Shoulder
    in arthritis
        juvenile chronic, 122
        rheumatoid, 114–115
        rheumatoid, radiogra-
            phy of, 115
    in calcium hydroxyapatite
        deposition disease,
        173
    in CPPD crystal deposi-
        tion disease, 168
    dislocation (*see* Disloca-
        tion, shoulder)
    impingement syndrome,
        231–232
    joint anatomy, 227–228
    Milwaukee, 173
    neuroarthropathy, 160
    normal, 228
    pseudosubluxation, 232
    trauma, 225–233
        key concepts, 225
Shunt aneurysm: in renal
    osteodystrophy, 308
Sickle cell, 382
    anemia, 381
    dactylitis and periosteal
        reaction in infant,
        398
    disease and avascular ne-
        crosis, 183
    thalassemia, 382
Sinding-Larsen-Johanssen
    disease, 179

Sinuses: pneumatized, in
    acromegaly, 311
Sjögren's syndrome, 119
Skeletal hyperostosis (*see*
    Hyperostosis, diffuse
    idiopathic skeletal)
Skin
    bronze, with cirrhosis
        and diabetes, 169
    hyperelastic, in Ehlers-
        Danlos syndrome,
        329
Skull
    in anemia, 380
    in dwarfism, thanato-
        phoric, 360
    in neurofibromatosis, 326
Smith fracture, 206
Sodium urate: in gout, 161
Soft tissue
    in bone tumors, 3
    calcification, 399–400
        collagen vascular dis-
            ease, 399–400
        congenital, 400
        drugs and, 400
        infections, 400
        key concepts, 399
        metabolic, 400
        trauma and, 399
        tumors and, 399
    fat planes in osteomyeli-
        tis, 384
    lipoma, 81
    osteosarcoma, 40
    sarcoma, surgical staging,
        6
    tumors, fibrous, 78–79
Spastic flatfoot (*see* Tarsal
    coalition)
Spina ventosa, 388
Spine
    in alkaptonuria, 174

Spine (*cont.*)
  in arthritis
    juvenile chronic, 123
    psoriatic, 133–134
    rheumatoid, 116–119
    "bamboo," 128
  bifid process, 273
  cervical, 264–276
    anomalies, congenital,
      269–273
    AP film, 267
    AP view, 268
    CT of, 269
    flexion injuries, 274–
      275
    fracture, bilateral verti-
      cal, 273
    fracture in children,
      276
    injury pattern of, 273–
      275
    lateral, with fanning of
      spinous processes,
      266
    lateral flexion-extension
      view, 268–269
    normal anatomy, 264–
      269
    normal position of
      bony elements, 264–
      265
    normal variants, 269–
      273
    oblique film, 267
    open-mouth film, 267
    ossification, 270–271
    ossification, normal
      centers, 270–271
    pillar views, 269
    radiographic signs of
      instability, 275
  in dwarfism, diastrophic,
    364

  in hyperostosis, diffuse
    idiopathic skeletal,
    186–187
  infection, 391–393
    anatomy, pertinent,
      391
    radiographic appear-
      ance, 392
  instrumentation in total
    knee arthroplasty,
    291
  lumbar, 276–279
    in achondroplasia, 357
    CT appearance of fac-
      ets, 278
    in dwarfism, thanato-
      phoric, 360
    fracture dislocation,
      277
    in hypochondroplasia,
      358
    normal appearance,
      276
    patterns of injury, 277–
      278
  neuroarthropathy, 160–
    161
  ossification in achondro-
    genesis, 361
  in osteoarthritis, 156–157
  osteoblastoma of, 23
  in osteomyelitis, 384
  in Paget's disease of
    bone, 316
  in Reiter's disease, 138
  in short stature, dispro-
    portionate, 373
  in spondylitis, 125, 126–
    128
  stenosis, 157
  thoracic, 276
    fracture, 276
  trauma, 264–279
    key concepts, 264

Spine (*cont.*)
  tuberculosis of, 392
  uncinate process, 273
Spiral fracture, 191
Spleen: infarction in ane-
    mia, 381
Spondylitis, 124–130
  ankylosing, 124–130
    joints commonly in-
        volved in, 127
    juvenile, 121
    juvenile, misdiagnosis
        of, 121
  definition, 125
  differential diagnosis,
    129–130
  epidemiology, 125
  extra-articular manifesta-
    tions, 125
  glenohumeral joint in,
    128
  hip in, 128
  joints commonly affected
    in, 126
  key concepts, 124
  laboratory tests in, 125
  radiography of, 125–126
    survey films for, 130
  signs, clinical, 125
  spine in, 126–128
  symmetry in, 129
Spondyloarthropathy: in
    renal osteodystro-
    phy, 308–309
Spondyloepiphyseal dys-
    plasia congenita,
    367–368
  clinical features, 367
  diagnosis, differential,
    368
  inheritance, 367
  prognosis, 368
  progression, 368
  radiographic features,
    367–368

Spondylolisthesis, 157
Spondylolysis
  in Ehlers-Danlos syn-
    drome, 329
  in Marfan's syndrome,
    327
Spondylosis, 278–279
  deformans, 156–157
Staging: surgical, 6
Staphylococcus
  in anemia, 381
  aureus, 382
    in discitis, 392–393
    in spinal infection, 391
  in drug abuse, 382
  sickle cell and, 382
Stature (*see* Short stature)
Stenosis: spine, 157
Sternoclavicular
  joint, 225–226
  septic arthritis, 391
Steroids
  in avascular necrosis, 183
  bone abnormalities due
    to, 317
  exogenous, 295–296
Still's disease, 120
Streptococcus: group B, 382
Stress fracture, 193
  ankle, 258
  differential diagnosis, 22
  of femur, 245
  fibula, 250
  metatarsal, 263
  in Paget's disease of
    bone, 316
  subacute, in osteomyeli-
    tis, 386
Subchondral
  cyst (*see* Cyst,
    subchondral)
  resorption in hyperpara-
    thyroidism, 305

Subchondral (*cont.*)
  sclerosis in osteoarthritis,
    150–151
Subligamentous resorption:
    in hyperparathyroid-
    ism, 305
Subluxation, 193
  carpal, dorsal, 216
Subperiosteal hemorrhage:
    in scurvy, 312
Subtalar joint: Harris-Beath
    view, 352
Sudeck's atrophy, 296
Supracondylar fracture, 222
Surgery treatment options
    for tumors, 8–9
Surgical resection: guidance
    of, 2
Swanson arthroplasty, 291
Symphysis pubis, 235

Synovial
  cell (*see* Sarcoma, synov-
    ial cell)
  chondromatosis (*see*
    Chondromatosis,
    synovial)
Synovitis
  pigmented villonodular,
    82, 175–176
    definition, 175
    extra-articular manifes-
      tations, 175
    key concepts, 175
    radiography of, 175–
      176
    transient, in septic arthri-
      tis of hip in children,
      390
Syphilis osteomyelitis, 388
Syringomyelia, 160

T

Talar dome osteochondritis
    dissecans, 180
Talipes equinovarus (*see*
    Clubfoot)
Talocalcaneal coalition,
    350–352
  CT of, 353
  Harris-Beath view, 352
Talocalcaneal joint, 262
Talonavicular joint: disloca-
    tion, 348
Talus
  congenital vertical, 346–
    347, 348
    radiographic findings,
      347
  trauma, 262–263
Tarsal coalition, 349–354
  arthrography, 353–354
  bone imaging in, 352
  CT in, 352–353
  epidemiology, 349–350

  key concepts, 349
  radiographic findings,
    350–352
  signs and symptoms, 350
  special techniques, 352–
    354
Tarsal navicular, 260
Tarsometatarsal joints: nor-
    mal alignment, 261
Tarsus adductus, 349
  radiographic findings,
    349
Teardrop burst fracture, 275
Technetium-99m
  -MDP gallium-67 WBC
    imaging in osteo-
    myelitis, 387
  in tumor work-up, 11
Telangiectatic (*see* Osteosar-
    coma, telangiectatic)

Temporomandibular joint:
  in juvenile chronic
    arthritis, 123
Tendon(s)
  in acromegaly, 312
  sheath, giant cell tumor
    of, 82
Tenotomy: adductor, in hip
    dislocation, congeni-
    tal, 335
Thalassemia, 381–382
  sickle cell, 382
Thanatophoric (*see* Dwarf-
    ism, thanatophoric)
Thoracic
  cage in asphyxiating tho-
    racic dysplasia, 362
  dysplasia (*see* Dysplasia,
    asphyxiating
    thoracic)
Thoracolumbar junction: in
    tuberculosis of spine,
    392
Thorax
  barrel-shaped, in achon-
    drogenesis, 361
  in chondroectodermal
    dysplasia, 363
Thrombotic episodes: in
    homocystinuria, 328
Thumb, 200–201
Thyroid
  acropachy, 311
    differential diagnosis,
      182
    periosteal reaction in,
      311
  disease, 310–311
Tibia
  apophysis in osteochon-
    drosis, 179
  chondroblastoma, 48
  in chondroectodermal
    dysplasia, 363

  epiphysis, fusion, 258
  fibroma of, ossifying, 26
  fracture (*see* Fracture,
    tibia)
  in neurofibromatosis, 326
  in Paget's disease of
    bone, 316
  tubercle apophysis, ante-
    rior, 249–250
  tumors of, 4
Tillaux' fracture: juvenile,
    259
Tomography, computed
  in hip dislocation, con-
    genital, 335
  in osteoporosis, 297
  of pelvic fracture, 235–
    237
  in septic arthritis
    of sacroiliac joint, 391
    sternoclavicular, 391
  of spine
    cervical, 269
    lumbar, appearance of
      facets, 278
  talocalcaneal coalition,
    353
  in tarsal coalition, 352–
    353
  in tumor work-up, 11–12
Trabeculae
  in anemia, 380
  in Paget's disease of
    bone, 315
  resorption in hyperpara-
    thyroidism, 305
Transcondylar fracture, 223
Transverse fracture, 191
Trauma, 190–291
  ankle (*see* Ankle, trauma)
  calcaneus, 262
  carpus, 206–216
  elbow, 218–224
    key concepts, 218
  foot, 260–264

Trauma (*cont.*)
  key concepts, 260
  generalizations, 190
  hand, 196–203
  hip, 239–247
    key concepts, 239
  knee, 247–254
    key concepts, 248
  navicular, 263
  patella, 251
  pelvis, 233–239
    key concepts, 233
  periostitis after, 282–283
  scoliosis and, 322
  shoulder, 225–233
    key concepts, 225
  soft tissue calcification
    and, 399
  spine, 264–279
    key concepts, 264
  talus, 262–263
  wrist, 203–218
Trevor-Fairbank disease,
  46–47
Triquetrum
  fracture, 211
  -hamate instability, 214–
    216
Trisomy
  18, 21, 355
Trochanter: lesser, avulsion
  fracture of, 244
Trochlear notch: in hemo-
  philia, 379
Tuberculosis, 387–388, 390
  of spine, 392
Tumor(s), 1–106
  age and, 2–3
  -associated rickets, 301–
    302
  bone (*see* Bone, tumors)
  brown (*see* Brown
    tumors)

cartilage-forming (*see* Car-
    tilage-forming
    tumors)
connective tissue, 78–86
  benign, 78–82
desmoid, 79
in diaphysis, 4
of epiphysis, 4
giant cell (*see* Giant cell
    tumors)
glomus, 74
host response and, 5
location of lesion, 4
margin of lesion, 5
marrow, 62–71
  WHO classification
    with modification, 15
of metaphysis, 4
monostotic, 5
in Paget's disease of
    bone, 316–317
periosteal response to,
    397–398
polyostotic, 5
radiation-induced, 394–
    395
scoliosis and, 322–323
size of lesion, 4
soft tissue calcification
    and, 399
soft tissue, fibrous, 78–79
surgical treatment op-
    tions, 8–9
of tibia, 4
treatment, complications
    of, 13–14
vascular, 71–78
  benign, 71–74
  intermediate or inde-
    terminate for malig-
    nancy, 74–75
  malignant, 76–78
  WHO classification
    with modification,
    15–16

Tumor(s) (*cont.*)
  visible matrix, 5
  work-up
    angiography in, 13
    biopsy in, 10
    CT in, 11–12
    MRI in, 12

Ulna
  deviation, 208
  dislocation in dyschon-
      drosteosis, 367
  -minus variant, 204
  normal tilt of distal radial
      articular surface on
      PA film, 204

Valgus
  abductus, 339
  angulation: of fracture,
      193
  forefoot (*see* Forefoot,
      valgus)
  hindfoot (*see* Hindfoot,
      valgus)
Varus
  adductus, 338
  angulation: of fracture,
      193
  forefoot (*see* Forefoot,
      varus)
  hindfoot (*see* Hindfoot,
      varus)
Vasoconstriction: in ane-
    mia, 380
Veins of Batson: in spinal
    infection, 391
Vertebra
  limbus, 156
  in pseudoachondroplasia,
      359
Vertebral body

  philosophy of, 9–17
  radiography in, 11
  radionuclide studies in,
      11
Tumoral calcinosis: differ-
    ential diagnosis, 283–
    284
Turner's syndrome, 355

U

  -plus variant, 204
  translocation, 216
Ultrasound: in hip disloca-
    tion, congenital, 335
Union: delayed, 193

V

  flattening in spondyloepi-
      physeal dysplasia
      congenita, 367
  infarct in anemia, 381
  necrosis in anemia, 381
  ossification in chondro-
      dysplasia calcificans
      punctata, 365
  scalloping
    in achondroplasia, 357
    in Ehlers-Danlos syn-
        drome, 329
    in homocystinuria, 328
    in Marfan's syndrome,
        327
    in neurofibromatosis,
        325
Vessels
  collagen, diseases, and
      soft tissue calcifica-
      tion, 399–400
  fragile, in Ehlers-Danlos
      syndrome, 329
  insufficiency, differential
      diagnosis, 182

Vessels (*cont.*)
  metastases (*see* Metastases, vascular)
  tumors (*see* Tumors, vascular)
Villonodular (*see* Synovitis, pigmented villonodular)
Vitamin
  A poisoning causing bone abnormalities, 317

D
  in osteomalacia, 299–300
  poisoning causing bone abnormalities, 317
  resistant rickets, 301–302
  in rickets, 299–300
Volar
  flexion of lunate, 216
  plate fracture, 199–200
von Willebrand's disease, 378

W

Wilson's disease, 170–171
  definition, 170
  differential diagnosis, 171
  epidemiology, 170
  extra-articular manifestations, 171
  joints most commonly affected, 171
  key concepts, 170
  radiography of, 171
  signs, clinical, 170
Wires: cerclage, in total knee arthroplasty, 290
Work-up (*see* Tumors, work-up)
Wormian bone
  in cleidocranial dysostosis, 354
  in hypothyroidism, juvenile, 310

  in osteogenesis imperfecta, 330
  in sclerosing dysplasia, 332
Wrist
  in arthritis
    juvenile chronic, 122
    rheumatoid, 112–113
  arthrography, 217–218
  in calcium hydroxyapatite deposition disease, 173
  in CPPD crystal deposition disease, 167
  in gout, 163
  lateral film normal anatomy, 204
  in osteoarthritis, 152, 153
  PA film normal anatomy, 203
  trauma, 203–218
    key concepts, 203